JN395250

# 세기의 과학적 비밀
Scientific secret of the century

# 세기의 과학적 비밀
Scientific secret of the century

**초판 1쇄 인쇄** 2024년 03월 08일
**초판 1쇄 발행** 2024년 03월 26일

**신고번호** 제313-2010-376호
**등록번호** 105-91-58839

**지은이** 윤종오

**발행처** 보민출판사
**발행인** 김국환
**기획** 김선희
**편집** 이상문
**디자인** 다인디자인

**ISBN** 979-11-6957-139-5 03510

**주소** 경기도 파주시 해올로 11, 우미린더퍼스트@ 상가 2동 109호
**전화** 070-8615-7449
**사이트** www.bominbook.com

한영합본

Korean-English combination

# 세기의 과학적 비밀

# Scientific secret of the century

윤종오 著

# 프롤로그
# Prologue

"기껏해야 과학은 (사회윤리적) 목적을 이루는 도구를 제시할 뿐이다. … 우리는 인간의 문제에 관한 한 과학과 과학적 방법을 과대평가해서는 안 된다. 또 우리는 사회조직에 영향을 미치는 문제에 대해 의사표시를 할 수 있는 사람들이 단지 전문가뿐이라는 생각을 해서도 안 된다."

"At best, science only presents tools to achieve (social and ethical) goals. … We should not overestimate science and the scientific method when it comes to human problems. Nor should we assume, that only experts can express themselves on matters affecting the social fabric."

– 알버트 아인슈타인(Albert Einstein)

이 책을 집필하고 출판하기까지 많은 고민이 있었다.

There were many concerns before writing this book and publishing it.

이 비밀들이 밝혀졌을 때, 세상 사람들이 이 비밀이 진실인지 아닌지 진지하게 실험해보고 토의해볼 것인지 알 수 없었기 때문에 가지는 고민이었다.

When these secrets were revealed, it was a concern because it was not known whether people in the world would seriously test and discuss whether these secrets were true or not.

이 책의 내용에 대한 진지한 접근 없이 시작될 수 있는 비판도 두려웠다.

I was also afraid of criticism that might begin without a serious approach to the contents of this book.

전문가들이 해결해줄 것이라 믿고 수년을 기다려보기도 했다.

I even waited for several years, believing that experts would solve the problem.

그러나, 이제 스스로 해결하지 않는 한 그 누구도 우리의 문제에 진정성을 가지고 접근하지 않는다는 것을 알게 되었다.

However, we now know that no one will approach our problems with sincerity unless we solve them ourselves.

제1부에서는 아토피의 모든 비밀을 파헤친다.

In Part 1, we uncover all the secrets of atopy.

2022년 전 세계 아토피 피부염 환자의 수는 2억 2,300만 명가지 증가하였다. 아토피증후군 환자에 해당하는 천식, 비염, 결막염까지 합하면 그 숫자는 헤아릴 수 없이 많아진다.

In 2022, the number of atopic dermatitis patients worldwide increased to 223 million. If you include asthma, rhinitis, and conjunctivitis, which are atopic syndrome patients, the number becomes countless.

과연 인류는 아토피의 원인을 몰라서 못 고치는 것이었을까?

Could it be that mankind is unable to cure atopy because they do not know the cause?

아니면 전문가들이 실수를 하고 있는 것일까?

Or are the experts making a mistake?

아니면 고의?

Or intentional?

아토피성 피부염의 원인이 규명되고, 해결책이 나온다는 것은 모든 아토피성 피부염 환자가 동일한 방법을 사용하여 동일한 치료효과를 얻게 된다는 것을 의미한다.

Identifying the cause of atopic dermatitis and finding a solution means that all atopic dermatitis patients will receive the same treatment effect using the same method.

이는 다른 아토피증후군 환자들도 치료효과를 기대할 수 있게 한다.

This allows other atopic syndrome patients to also expect treatment effects.

국내외 제약회사 중 아토피의 원인을 밝히려는 제약회사는 없다. 단지 눈에 띄는 증세를 호전시키거나 유지할 약물을 연구할 뿐이다.

Among domestic and foreign pharmaceutical companies, no pharmaceutical company is trying to determine the cause of atopy. We simply study drugs that will improve or maintain noticeable symptoms.

원인을 못 밝히면서 완벽한 치료를 할 수 있는가?

Is it possible to provide a perfect treatment without being able to identify the cause?

듀피젠트의 성공은 치료하지 않고 증세만 호전시키는 약품의 성공이었다.

The success of Dupixent was the success of a drug that only improved symptoms without treating the disease.

그렇게 대형 제약사들은 듀피젠트의 성공 방정식을 따라갈 뿐 아토피의 원인을 찾아내는 데에는 시간을 할애하지 않았다.

In this way, large pharmaceutical companies only followed Dupixent's formula for success and did not take the time to find the cause of atopy.

이제 더 이상 전문가들이 아토피를 치료하는 방법을 찾을 수 있으리라는 믿음이 없어질 무렵 새로운 막수송체를 발견했다.

Just when experts no longer had faith that they could find a way to treat atopy, a new membrane transporter was discovered.

새로운 막수송체는 아토피의 원인을 피부 밖으로 인도하여 인간이 관찰할 수 있도록 했다.

The new membrane transporter guides the cause of atopy outside the skin so that it can be observed by humans.

그리고, 치료용 세포 성장인자와 소독제를 세포 속으로 수송하여 아토피가 치료될 수 있음을 보여줬다.

Additionally, they showed that atopic dermatitis can be treated by transporting therapeutic cell growth factors and disinfectants into cells.

제2부에서는 '기초물리학의 흠결'에 대하여 다룬다.

Part 2, deals with 'flaws in basic physics.'

2023년 인류는 심각한 기후 변화의 중심에 서기 시작했다. 그런데도 인류는 아직 에너지 문제를 해결하지 못했다.

In 2023, humanity begins to find itself at the center of severe climate change. However, humanity has not yet solved the energy problem.

생계를 위한 에너지의 자유도 얻지 못했고, 에너지를 기후 변화 대응과 환경보호를 위해 사용할 준비가 되어있지 않다.

We have not achieved freedom of energy for livelihood and are not prepared to use energy to respond to climate change and protect the environment.

과연 인류는 에너지를 소비하는 종족에서 창조하는 종족으로 진화하지 못하는 것일까?

Is it true that humanity cannot evolve from a species that consumes energy to a species that creates energy?

에너지 보존의 법칙은 영원히 깨지지 않는 법칙일까?

Is the law of conservation of energy unbreakable?

에너지가 부족해서 지구의 기후 변화에 대응하지 못한 채로 지구의 환경을 관리하지 못하는 존재로 끝나는 것일까?

Will we end up unable to manage the Earth's environment due to lack of energy and unable to respond to climate change?

부족한 에너지를 더 가지기 위해서 끊임없이 싸워야 하는 종족으로 인류는 기억될까?

Will humanity be remembered as a species that must constantly fight to have more energy?

기초물리학은 에너지와 어떤 연관성이 있을까?

How does basic physics relate to energy?

**'아인슈타인이 틀렸다고 말할 수 있는 용기'**
**'The courage to say that Einstein was wrong'**

**'뉴턴이 틀렸다고 말할 수 있는 용기'**
**'The courage to say that Newton was wrong'**

인류에게 그런 용기와 지혜가 필요한 때다.
This is a time when humanity needs such courage and wisdom.

이 글을 작성하고 용기를 내도록 도움 주신 많은 분들께 감사드린다.
I am grateful to the many people who helped me write this article and find the courage to do so.

2024년 3월
저자 **윤종오**

# 목차
# Contents



제1부

## 막수송체 발견으로 밝혀진 아토피의 진실

Part 1

## The truth about atopy revealed through the discovery of a new membrane transporte

제2부

# 기초물리학의 흠결

Part 2

# Flaws in basic physics

# 부록
# supplement

# 막수송체 발견으로 밝혀진 아토피의 진실

The truth about atopy revealed through the discovery of a new membrane transporte

아토피 못 고치는 것인가 방치하는 것인가?

Is atopy not curable or is it neglected?

아토피의 비밀이 폭로된다.

The secret of atopy is revealed.

## 아토피의 비밀을 밝히는 밑거름
## The foundation for revealing the secrets of atopy

한국에서 유기규소이온액이 개발되었다.

Organosilicon ionic liquid has already been developed and is commercially available in Korea.

이 개발된 유기규소이온액이 세포막수송체 역할을 한다는 것은 아주 우연히 밝혀졌다.

It was discovered quite by chance that this developed organosilicon ionic liquid acts as a cell membrane transporter.

지금까지 인류의 과학적 발견은 막수송체는 세포막을 통과하는 물질들을 수송하는 단백질들로 알려져 왔다.

Until now, mankind's scientific discoveries have known that membrane transporters are proteins that transport substances through the cell membrane.

막수송체에 대한 연구는 세포의 기본적인 작용 원리를 이해하고, 질병의 원인과 치료법을 찾고, 새로운 약물을 개발하는 데에

크게 기여할 수 있는 학문으로 발전하여 막수송체학을 이루어 세포생물학의 중요한 분야로 발전하였다.

The study of membrane transporters has developed into a discipline that can greatly contribute to understanding the basic operating principles of cells, finding causes and treatments of diseases, and developing new drugs, forming membrane transportomics and becoming an important field of cell biology. developed.

개발된 킬레이트 방식의 규소이온액이 막수송체 역할을 한다는 것이 밝혀짐으로 인하여 인류는 세포 속으로 여러 가지 물질을 수송할 수 있는 길이 열렸다고 할 수 있다.

It can be said that the discovery that the developed chelate-type silicon ionic liquid acts as a membrane transporter has opened the way for mankind to transport various substances into cells.

실리콘이온이 막수송체로 작용한다는 개념은 이미 확립된 세포생물학 원리에서 벗어난 것으로 기존 학계에서 받아들이기 어려울 것이다.

The concept that silicon ions act as membrane transporters deviates from already established principles of cell biology and will be difficult for the existing academic community to accept.

전통적인 세포생물학에서 막수송체는 지금까지는 단백질로 알려졌었기 때문이다.

This is because in traditional cell biology, membrane transporters were known as proteins until now.

막수송체로서 실리콘이온의 역할을 제안하는 최근 개발이나 새로운 발견이 있었다면 이는 세포과정에 대한 인류의 이해에 중요한 변화를 의미한다.

If there were any recent developments or new discoveries suggesting a role for silicon ions as membrane transporters, this would represent a significant change in humanity's understanding of cellular processes.

실리콘이온성 액체의 개발과 막수송체에서의 잠재적인 역할에 대한 탐구는 과학 연구에 있어 흥미로운 일이다.

The development of silicon ionic liquid and the exploration of its potential role in membrane transporters are interesting aspects of scientific research.

유기규소이온액이 막수송체 역할을 한다는 것이 사실이라면, 세포생물학 분야에서는 혁명적인 일이다.

If it is true that organosilicon ionic liquids act as membrane transporters, it is revolutionary in the field of cell biology.

### 1. 막수송 기능

### 1. Membrane Transport Function

**2. 다양한 응용**

**2. Versatility and Applications**

**3. 생체적합성 및 안전성**

**3. Biocompatibility and Safety**

**4. 약물 전달에 미치는 영향**

**4. Impact on Drug Delivery**

**5. 작용 메커니즘**

**5. Mechanisms of Action**

**6. 생화학 및 생물물리학 연구**

**6. Biochemical and Biophysical Studies**

등 앞으로 많은 연구가 필요할 것이다.
Much research will be needed in the future.

유기규소이온액은 이미 개발되어 한국에서 시판되고 있다.
Organosilicon ionic liquid has already been developed and is commercially available in Korea.

해당 유기규소이온액에는 그 어떠한 단백질 종류도 들어있지 않으므로 누구나 쉽게 구매하여 실험을 통하여 증명할 수 있다. (유기규소이온액이 막수송체 역할을 하는지 여부를 확인할 수 있다.)

Since the organosilicon ionic liquid does not contain any type of protein, anyone can easily purchase it and prove it through experiments. (You can check whether the organosilicon ionic liquid acts as a membrane transporter.)

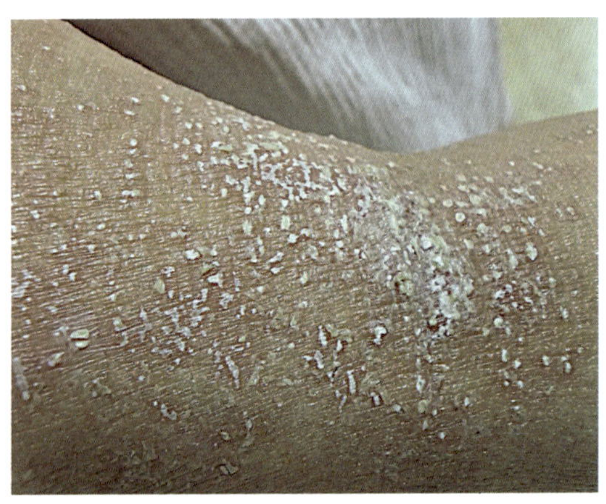

규소이온 기반의 소독용 화장품으로 아토피 피부염 환자의 아토피 부위를 1시간 이상 긴 시간 소독할 경우 아토피 피부염 원인균을 만날 수 있다. (해산의 고통에 버금가는 고통이 따른다.)

If you disinfect the atopic area of an atopic dermatitis patient for more than an hour with silicon ion-based disinfecting cosmetics, you may encounter bacteria that cause atopic dermatitis. (There is pain comparable to the pain of childbirth.)

인간이 사용하기 쉬운 막수송체는 세포에 소독제를 전달하여 사진처럼 원인균을 소독하여 피부 바깥에서 관찰할 수 있는 기회를 제공한다.

Membrane transporters, which are easy for humans to use, deliver disinfectants to cells, disinfecting causative bacteria as shown in the photo, and providing an opportunity to observe them from outside the skin.

킬레이트 방식의 유기규소이온액이 막수송체 역할을 한다는 것이 밝혀지므로 인하여, 아토피의 원인균이 세포 내 기생충임을 밝혀내고 치료할 수 있는 길이 열린 것이다.

With the discovery that chelating organosilicon ion liquid acts as a membrane transporter, the path to treatment has been opened by revealing that the causative agent of atopy is an intracellular parasite.

이 유기규소이온액 제조기술은 한국, 호주, 중국, 인도, 캐나다 등 5개국에 특허가 등록되었고, 미국, 일본, 유럽연합(EU)에 특허 심사 중에 있다.

This organosilicon ionic liquid manufacturing technology has been patented in five countries, including Korea, Australia, China, India, and Canada, and is under patent examination in the United States, Japan, and the EU.

또한 아토피 피부염 치료 관련 특허는 현재 한국에 출원되어 있다.

Additionally, a patent related to the treatment of atopic dermatitis is currently applied for in Korea.

이 유기규소이온이 아토피 치료에서는 다음과 같이 작동한다.

This organosilicon ion works as follows in the treatment of atopic dermatitis.

**1. 킬레이트 실리콘 이온액의 섭취는 몸 속에 숨어있는 아토피 원인균을 세포에서 쫓아내어 피부로 이동시키는 역할을 한다.**

**1. Consumption of chelating silicon ionic liquid plays the role of expelling atopy-causing bacteria hidden in the body from cells and moving them to the skin.**

**2. 실리콘이온이 이온채널을 여는 역할을 한다.**

**2. Silicon ions play the role of opening ion channels.**

**3. 소독제 및 세포 성장인자를 운송한다.**

**3. Transport disinfectants and growth factors.**

**4. 전체적인 목적은 치료제의 세포 내로의 전달을 가능하게 하여 피부질환 치료용 조성물을 구축한다.**

**4. The overall purpose is to construct a composition for treating skin diseases by enabling delivery of the therapeutic agent into cells.**

따라서 막수송체 기능 여부는 누구나 쉽게 구매하여 실험이 가능하다. (굳이 갑론을박할 필요가 없다.)

Therefore, anyone can easily purchase and test whether the membrane transporter functions. (There is no need to argue.)

규소이온 500g + EGF 1ppm 10ml + FGF 1ppm 10ml + IGF 1ppm 10ml를 희석한 제품과 성장인자만 단독으로 사용했을 때를 비교 분석하는 것만으로도 규소이온의 막수송체 역할을 검증할 수 있다.

The role of silicon ions as a membrane transporter can be verified simply by comparing and analyzing the diluted product of 500g of silicon ions + 10ml of EGF 1ppm + 10ml of FGF 1ppm + 10ml of IGF 1ppm and when the growth factors are used alone.

이로써 아토피 원인에 대한 불확실성을 종식시키고, 치료방법을 세상에 제시한다.

This puts an end to uncertainty about the cause of atopy and presents a treatment method to the world.

그러나 치료방법이 나왔다고 모든 것이 끝난 것은 아니다. 치료에는 균의 사멸로 인한 엔도톡신 반응에 의한 고통이 따른다.

However, just because a treatment method is found does not mean everything is over. Treatment involves pain due to the endotoxin reaction caused by the death of bacteria.

그 외에 효율적인 치료방법에 대한 추가 연구가 필요할 것이다.

Additionally, additional research will be needed on effective treatment methods.

이 책이 발간되면 이 분야 전문가들이 연구하지 않을까?

When this book is published, won't experts in this field research it?

**관련 제품**
**(www.si-ion.com)**

**유황이온액**
**sulfur ionic liquid**

ICID Application No. 2023-10732

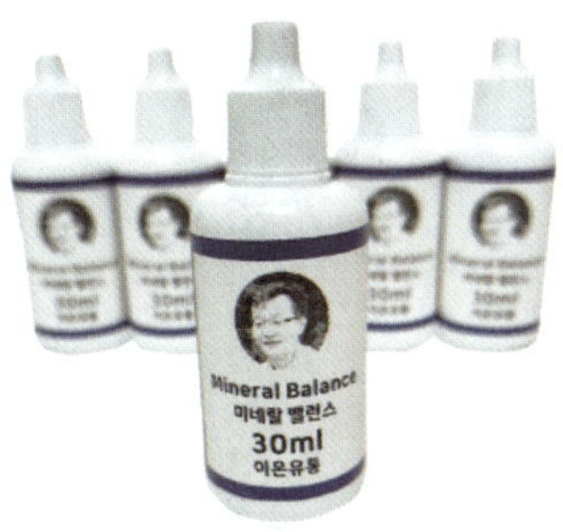


**규소이온 기반의 미네랄 밸런스**
**Silicon ion-based mineral balance**

이중이온 방식의 섭취용 종합 미네랄 이온액
Mineral ionic liquid for general consumption manufactured using the double ion method

**유기규소이온액 K9**

ICID Application No. 4-07-2021-12380

**유기규소이온액 K8**

섭취용 규소 미네랄 영양소

silicon mineral nutrients

# 아토피 증후군
## atopic syndrome

아토피(atopy)라는 용어는 Coca와 Cooke가 1923년에 만들어낸 용어이다. 2023년은 이 용어가 만들어진 지 101년이 된 해이다.

The term atopy was coined by Coca and Cooke in 1923. 2023 marks the 101st year since this term was coined.

아토피 또는 아토피증후군은 알레르기 항원에 대한 직접 접촉 없이 신체가 극도로 민감해지는 알레르기 반응을 말한다.

Atopy or atopic syndrome refers to an allergic reaction in which the body becomes extremely sensitive to allergens without direct contact.

아토피증후군 증상으로는 (1) 아토피 피부염, (2) 알레르기성 결막염, (3) 알레르기성 비염, 그리고 (4) 천식이 있다.

Symptoms of atopic syndrome include (1) atopic dermatitis, (2) allergic conjunctivitis, (3) allergic rhinitis, and (4) asthma.

101년이란 아토피의 역사는 아이러니하게도 아토피의 원인이 밝혀져도 서로 믿지 못하는 사회현상을 발생시켰다.

Ironically, the 101-year history of atopy has created a social phenomenon in which people do not trust each other even when the cause of atopy is revealed.

지난 100여 년 동안 인간은 세뇌되었고, 잘못 알려진 아토피의 원인이 머릿속에 각인되어 버렸다.

Over the past 100 years, humans have been brainwashed, and the poorly known causes of atopy have been imprinted in our minds.

새로운 명확한 아토피의 원인이 밝혀져도 기존 아토피의 원인을 믿었던 사람들과, 아토피의 원인이 불분명하다고 믿어왔던 사람들이 반발한다.

Even if a new clear cause of atopy is revealed, people who believed in the existing cause of atopy and people who believed that the cause of atopy was unclear protest.

이 책 속의 아토피 원인은 반발할 이유가 없다. 아토피의 원인과 치료방법을 공개하기 때문이다.

There is no reason to object to the causes of atopy in this book. This is because it discloses the causes and treatment methods of atopy.

따라서, 실천해보면 이 책 속의 아토피의 원인을 스스로 믿게 된다.

Therefore, if you practice it, you will yourself believe in the causes of atopy in this book.

그러나 실천하지 않는 사람들의 투정은 어쩔 수 없다.

However, the complaints of those who do not practice are inevitable.

필자는 과학하는 사람으로서 과학적인 방법의 원인과 과정 그리고 결과를 내놓는 것이다.

As a scientist, I present the causes, processes, and results of the scientific method.

오직 실천하여 아토피를 치료하는 것은 실천하는 자들의 몫이다.

It is the responsibility of those who practice to treat atopy only through practice.

실천하지 않고, 도전하지 않는 사람들의 푸념은 절대로 현실을 변화시키지 못하고, 자기 자신을 변화시키지 못한다.

The complaints of people who do not practice and do not challenge can never change reality or themselves.

분명한 점은 이 책으로 인하여, 아토피 피부염은 통제 불가능한 질병에서 통제 가능한 질병으로 분류가 변할 것이라는 점이다.

What is clear is that with this book, the classification of atopic dermatitis will change from an uncontrollable disease to a controllable disease.

## 아토피 피부염
## Atopic dermatitis

대한민국 건강보험심사평가원이 발표한 자료에 따르면, 한국 내 아토피 피부염의 환자 수는 2021년 98만 7,000여 명이다. (이것도 비염이나 천식, 결막염 환자의 숫자는 제외한 것이다.)

According to data released by the Health Insurance Review and Assessment Service of the Republic of Korea, the number of atopic dermatitis patients in Korea is approximately 987,000 in 2021. (This also excludes the number of patients with rhinitis, asthma, and conjunctivitis.)

세계적으로는 2022년(GBD 2022)에 약 2억 2,300만 명이 아토피 피부염을 앓고 있으며, 이중 약 4,300만 명이 1~4세이다. (출처: Global Report on Atopic Dermatitis 2022)

Worldwide, approximately 223 million people are suffering from atopic dermatitis in 2022 (GBD 2022), of which approximately 43 million are aged 1 to 4. (Source: Global Report on Atopic Dermatitis 2022)

의학계에서는 아토피 피부염의 발생원인이 복잡하다고 보는 경향도 있고, 불분명하다고 보는 경향도 있다.

In the medical community, there is a tendency to view the cause of atopic dermatitis as complex and unclear.

대한민국 질병관리청 국가 건강정보 포털 내용을 중심으로 보면, 아토피 피부염의 발병원인에 대하여 "많은 연구가 진행되어 왔으나, 지금까지 그 원인은 완전히 밝혀지지 않았습니다. 아토피 피부염은 유전적인 소인과 환경적인 요인, 면역학적 이상과 피부 보호막 역할을 하는 피부장벽 기능의 이상 등 요인들의 복잡한 상호 작용에 의해 발생하는 것으로 생각되고 있습니다."라고 밝혀 원인불명임을 명확히 인정하고 있다.

According to the contents of the National Health Information Portal of the Korea Disease Control and Prevention Agency, "A lot of research has been conducted on the cause of atopic dermatitis, but the cause has not been fully revealed until now. Atopic dermatitis is thought to be caused by a complex interaction of factors such as genetic predisposition, environmental factors, immunological abnormalities, and abnormalities in the skin barrier function that acts as a skin barrier." he said, clearly acknowledging that the cause is unknown.

한동안 의학계는 아토피 피부염의 발병 원인을 인체 내에 기생충이 너무 없어 기생충을 잡을 때 작동하던 면역체계에 해당하는 IL-4, IL-13이 일으키는 면역 과잉반응이라는 가설을 세웠다.

For a while, the medical community hypothesized that the cause of atopic dermatitis was an immune overreaction caused by IL-4 and IL-13, which correspond to the immune system that operates when catching parasites because there are too many parasites in the human body.

그러나, 인체에는 다양한 종류의 상재균을 포함하는 기생충들이 존재하고 있는 것이 밝혀졌다.

However, it has been revealed that parasites, including various types of commensal bacteria, exist in the human body.

그럼에도 불구하고 면역 과잉반응 가설은 듀피젠트라는 아토피 신약이 개발되는 토대가 되기도 하였다.

Nevertheless, the immune overreaction hypothesis also served as the basis for the development of a new atopic drug called Dupixent.

듀피젠트는 IL-4와 IL-13의 신호전달을 인위적으로 억제하는 방식의 대증요법 또는 대증치료(Symptomatic treatment)용 의약품인 것이다.

Dupixent is a symptomatic or symptomatic treatment medicine that artificially suppresses signaling of IL-4 and IL-13.

대증치료는 질병의 원인을 치료하는 것이 아닌, 질병의 증상만을 치료하는 의료방법 중 하나이다.

Symptomatic treatment is a medical method that treats only the symptoms of a disease, rather than treating the cause of the disease.

즉, 근본적인 치료가 되지 못할 때, 또는 근본적인 치료방법이 없거나 아직 치료방법을 찾아내지 못했을 때, 질병의 현상과 증상을 감소시킴으로 환자에게 위로와 평안을 주는 역할을 한다.

In other words, when fundamental treatment is not possible, or when there is no fundamental treatment method or a treatment method has not yet been found, it serves to provide comfort and peace to the patient by reducing the symptoms and symptoms of the disease.

이렇듯 아토피 피부염은 아직까지 발병의 원인조차 제대로 밝혀내지 못하고 있는 실정이었다.

As such, the cause of atopic dermatitis has not yet been properly identified.

# 알레르기성 비염
Allergic rhinitis

비염은 알레르기성 비염과 비알레르기성 비염으로 나눌 수 있는데 비염의 원인 또한 아토피 피부염처럼 그 원인이 명확하지 않다.

Rhinitis can be divided into allergic rhinitis and non-allergic rhinitis, and like atopic dermatitis, the cause of rhinitis is not clear.

비염은 비강을 덮고 있는 점막의 염증성 질환을 의미하는데 해당 염증이 왜 생기는지 아직 명확하게 밝혀지지 않았기 때문이다.

Rhinitis refers to an inflammatory disease of the mucous membrane covering the nasal cavity, and it is not yet clear why the inflammation occurs.

비강은 아토피 피부염이 창궐하는 연약한 부위의 피부와 유사하게 외부의 공기가 접촉하는 연약한 부위이고, 비강점막은 코 안에 위치한 구조물로 하비갑개, 중비갑개, 상비갑개로 이루어져 있으며, 공기의 온도와 습도를 조절하는 역할을 담당하며 비강점막에 문제가 있으면 기관지 확장증과 천식, 비염이 발생하게 된다.

The nasal cavity is a fragile area that comes into contact with external air, similar to the skin in the vulnerable area where atopic dermatitis is prevalent. The nasal mucosa is a structure located inside the nose and consists of the inferior turbinate, middle turbinate, and superior turbinate, and regulates the temperature and humidity of the air. It plays a role in regulating the nasal mucosa, and if there is a problem with the nasal mucosa, bronchiectasis, asthma, and rhinitis occur.

비강점막은 붉고 촉촉하며, 부드럽고 연약한 조직으로 되어있다.

The nasal mucosa is red and moist and is made of soft and fragile tissue.

피부의 연약한 부위보다 더 연약하며 외부공기와 직접적인 접촉을 하는 신체 부위로 비염은 아토피 피부염과도 상당한 연관성을 가지고 있다.

Rhinitis is a part of the body that is more fragile than the delicate skin and is in direct contact with the outside air and has a significant relationship with atopic dermatitis.

위키백과에 기술된 바와 같이 알레르기성 비염 역시 아토피증후군 증상에 해당하기 때문이다.

As described in Wikipedia, allergic rhinitis is also a

symptom of atopic syndrome.

한편, 염증반응(inflammatory responses)은 미생물의 감염이나 외부의 손상으로부터 생체를 보호하기 위한 방어기전으로 이러한 염증반응은 비교적 최근에 많은 발전이 이루어진 면역학의 발전과 더불어 선천면역이라는 개념과 질병과의 연계로 인하여 조명받고 있다.

Meanwhile, inflammatory responses are a defense mechanism to protect the living body from microbial infection or external damage. These inflammatory responses are linked to the concept of innate immunity and disease, along with the recent development of immunology. It is in the spotlight because of this.

아토피 피부염과 비염은 인체의 최 외각에 위치하는 연약한 피부와 비강점막에 나타나는 다양한 염증반응에 해당하는 현상이며, 이 염증반응을 발생시키는 원인에 해당하는 주변환경의 변화를 해석하지 못해 치료에 난항을 겪고 있는 것이다.

Atopic dermatitis and rhinitis are phenomena that correspond to various inflammatory reactions that occur in the delicate skin and nasal mucosa located in the outermost layer of the human body, and treatment is difficult due to the inability to interpret changes in the surrounding environment that cause this inflammatory reaction. are experiencing it.

일반적으로 아토피 피부염은 사람이 가려운 부위를 긁어서 염증을 일으키는 것으로 관찰자가 잘못 해석할 수 있다.

In general, observers may misinterpret atopic dermatitis as a condition that causes inflammation due to a person scratching an itchy area.

그러나, 가려움증 자체가 세포에서 발생한 고통현상이며, 따라서 가려움증은 사람이 건드리지 않아도 발생하는 비염과 같은 염증반응이다.

However, itching itself is a painful phenomenon that occurs at the cellular level, and therefore, itching is an inflammatory reaction like rhinitis that occurs even without a person touching the skin.

대표적으로 아토피 피부염과 비염을 예로 들었으나, 알레르기성 결막염, 천식 및 아직 원인이 밝혀지지 않은 많은 종류의 피부질환(예를 들면, 습진, 화폐상 습진, 현미경으로 관찰되지 않는 원인으로 발생하는 피부질환 등)도 비슷한 실정에 처해있다.

Atopic dermatitis and rhinitis are representative examples, but allergic conjunctivitis, asthma, and many types of skin diseases whose causes are not yet known (e.g., eczema, nummular eczema, skin diseases caused by causes that cannot be observed under a microscope etc.) are also in a similar situation.

# 아토피증후군의 자가면역질환 여부에 대한 고찰
# Consideration on whether atopy syndrome is an autoimmune disease

(논문) 말라리아 기생충 및 알레르기에 대한 3개월간 알벤다졸 치료의 효과: 가구 기반 클러스터 무자위 배정, 이중 맹검, 위약 대조 시험 (PMC3602425)

(Paper) Effect of 3-month albendazole treatment on malaria parasites and allergies: a household-based cluster random assignment, double-blind, placebo-controlled trial (PMC3602425)

이 논문으로 기생충을 치료하였음에도 불구하고 아토피가 증가하지 않는다는 것을 확인할 수 있었다. 그러므로 기생충의 치료는 아토피증후군의 증가에 영향을 미치지 않음을 확인할 수 있었다.

This paper confirmed that atopy did not increase despite treatment of parasites. Therefore, it was confirmed that treatment of parasites had no effect on the increase in atopic syndrome.

즉, 알벤다졸로 구충되는 기생충은 아토피와 관련이 없음을 알 수 있다는 것이다.

In other words, it can be seen that parasites dewormed with albendazole are not related to atopy.

아토피가 자가면역질환이 아니면 인간의 면역체계는 무엇인가와 싸우고 있는 것이고 그 싸움과정에서 염증반응이 생기는 것이다.

If atopy is not an autoimmune disease, the human immune system is fighting something, and an inflammatory response occurs during the fight.

가려움증은 세포에서 발생하는 염증반응에 의한 고통의 크기를 뇌에서 느꼈을 때, 가려움증으로 느낀 것이다.

Itching is felt when the brain feels the amount of pain caused by the inflammatory response that occurs at the cellular level.

아토피 피부염이 낫지 않고 염증반응이 계속되는 이유는 우리의 면역체계가 계속 싸움에서 지고 있는 것으로 해석할 수 있다.

The reason why atopic dermatitis does not get better and the inflammatory response continues can be interpreted as our immune system continuing to lose the fight.

이러한 현상들을 종합하면 아토피가 자가면역질환이 아니면, 인간의 면역체계는 아토피를 일으키는 확인되지 않은 존재와 지속적으로 싸우고 있다는 것을 뜻하고, 그 존재는 알벤다졸에 영향을 받아서 위축되는 경우는 있어도 알벤다졸로 구충할 수 없는 존재일 가능성이 높다.

Putting these phenomena together, it means that unless atopy is an autoimmune disease, the human immune system is continuously fighting against an unidentified entity that causes atopy, and that entity is unknown even if it is affected by albendazole and may atrophy. There is a high possibility that it cannot be dewormed with albendazole.

따라서 자가면역질환이 아니라면, 아토피증후군과 관련된 여러 가지 가설을 재검토할 필요가 있다.

Therefore, if it is not an autoimmune disease, there is a need to reexamine various hypotheses related to atopic syndrome.

그중 음모론을 제일 먼저 살펴보면 전 세계 아토피증후군 환자의 수가 음모론이 진실이 아님을 쉽게 입증하고 있다.

Among them, if we look at the conspiracy theory first, the number of atopic syndrome patients around the world easily proves that the conspiracy theory is not true.

WHO 발표에 의하면 전 세계 어린이 인구의 20%가 아토피라고 한다.

According to the WHO, 20% of the world's children have atopic dermatitis.

이것을 달리 해석하면, 전 세계의 수많은 의사들이 전부 음모에 가담하지 않는 한, 음모론 가설은 불가능에 가깝다는 이야기가 된다.

To interpret this differently, unless numerous doctors around the world are all involved in a conspiracy, the conspiracy theory hypothesis is close to impossible.

결국 아토피가 자가면역질환이 아니라면, 아토피 염증을 일으키는 존재는 흔하면서도 특별한 존재일 가능성이 높다.

Ultimately, if atopy is not an autoimmune disease, the entity that causes atopic inflammation is likely to be a common yet special entity.

따라서 그 존재의 조건을 들어보면 최소한 다섯 가지의 조건에 부합해야 한다.

Therefore, if we look at the conditions for its existence, it must meet at least five conditions.

아토피 염증을 일으키는 존재는,

첫째, 흔한 존재이면서도 수많은 의사들의 눈에 띄지 않는 존재여야 한다. 흔한 존재여야 전 세계 어린이 20% 이상을 감염시켜 아토피를 겪게 할 수 있기 때문이다

First, it must be a common entity but one that cannot be noticed by many doctors. This is because it must be common enough to infect more than 20% of children around the world and cause them to suffer from atopy.

둘째, 아토피 피부염, 알레르기성 비염, 천식, 알레르기성 결막염 등 아토피증후군 질병에 공통적으로 관여할 가능성이 있어야 한다.

Second, there must be a possibility of being commonly involved in atopic syndrome diseases such as atopic dermatitis, allergic rhinitis, asthma, and allergic conjunctivitis.

셋째, 현재까지는 질병의 원인으로 의심받지 않는 존재일 가능성이 높다.

Third, it is highly likely that it is an entity that has not yet been suspected as the cause of the disease.

넷째, 세포 내에 존재할 수 있는 존재여야 한다. (일반적인 방법으로 관찰하기 어렵고 발견돼도 치료가 어려운 존재여야 한다.)

Fourth, it must be something that can exist within cells. (It must be difficult to observe using general methods and difficult to treat even if discovered.)

다섯째, 인간 이외의 동물에서 유사한 사례가 존재하는지 여부이다.

Fifth, whether similar cases exist in animals other than humans.

이러한 다섯 가지의 조건에 부합하는 것은 마이코플라즈마(Mycoplasma) 미세기생충이 있다. '미코플라즈마'라고 불리기도 한다.

Mycoplasma is a microscopic parasite that meets these five conditions.

마이코플라즈마균의 존재는 세포벽이 없어 그 형태가 일정하지 않으며 인공배지에서 증식 가능한 가장 작은 세균, 바이러스와 세균의 중간 성질을 가진 미생물로서 호흡기 감염병 등을 유발시킨다.

Mycoplasma bacteria do not have a cell wall, so their form is not constant. They are the smallest bacteria that can grow in artificial media and are microorganisms with intermediate properties between viruses and bacteria, causing respiratory infectious diseases.

사람에게서 분리된 마이코플라즈마는 16종이 보고되고 있으며, 이중 병원성이 인정된 균종으로는 폐렴을 일으키는 M.pneumoniae, 비뇨생식기계 감염을 일으키는 M.hominis, M.genitalium, U.urealyticum 4종이 있으며 그 외 12종은 아직 밝혀진 바 없다.

16 types of mycoplasma isolated from humans have been reported, and among these, the recognized pathogenic types include M.pneumoniae, which causes pneumonia, M. hominis, M. genitalium, and U.urealyticum, which cause genitourinary tract infections, and others. 12 species have not yet been identified.

마이코플라즈마균은 땅에 존재하는 균으로서 식물을 감염시키고, 초식동물 또는 초식동물을 잡아먹은 육식동물에 기생한다. 크기는 보통 2~5μ 정도이며 50μ 이상의 것도 있다.

Mycoplasma bacteria are bacteria that exist on land, infect plants, and parasitize herbivores or carnivores that eat herbivores. The size is usually about 2~5 μ, but some are larger than 50 μ.

예전에는 바이러스로 다루어졌으며, 그 후에는 세균으로 다루고 있다.

In the past, it was treated as a virus, and later it was treated as a germ.

사람의 마이코플라즈마 감염은 17% 정도가 피부의 발진과 연관되어 있다. (James, William D.; Berger, Timothy G.; 외. (2006). 《Andrews' Diseases of the Skin: clinical Dermatology》. Saunders Elsevier. ISBN 978-0-7216-2921-6.)

Approximately 17% of human mycoplasma infections are associated with skin rashes. (James, William D.; Berger, Timothy G.; et al. (2006). Andrews' Diseases of the Skin: clinical Dermatology. Saunders Elsevier. ISBN 978-0-7216-2921-6.)

이렇게 흔한데 눈에 띄지 않은 이유는 광학현미경으로는 관찰할 수 없기 때문이다.

The reason it is so common but not noticeable is because it cannot be observed with an ordinary optical microscope.

일반 광학현미경으로는 세포 속에 들어있는 세포 내 기생충을 관찰할 수 없다.

Intracellular parasites contained within cells cannot be observed using a general light microscope.

또한, 마이코플라즈마는 사람, 포유류, 곤충류, 식물 등에서 각각 감염 양상이 다르며, 사람에게 병원성을 나타내는 균종 이외에는 무시당하고 있다.

In addition, mycoplasmas have different infection patterns in humans, mammals, insects, and plants, and are ignored except for fungal species that are pathogenic to humans.

즉, 수많은 의사들이 마이코플라즈마균의 존재를 알고 있지만, 병원성을 나타내는 또는 병원성이 확인된 균종 이외에는 무시하는 흔한 존재가 마이코플라즈마균인 것이다.

In other words, although many doctors are aware of the existence of mycoplasma bacteria, mycoplasma bacteria are common entities that are ignored other than those that are pathogenic or have confirmed pathogenicity.

마이코플라즈마는 총 100여 종 중 실제 폐렴을 유발하는 병원균을 포함하여 인체에서 발견된 것은 16종이다.

Out of a total of 100 types of mycoplasmas, 16 types have been found in the human body, including the pathogen that actually causes pneumonia.

그중 12종의 병원성은 알려지지 않았으며 연구가 불확실하다.

Among them, the pathogenicity of 12 species is unknown and research is uncertain.

12종은 감염돼도 증상을 잘 모르는 경우가 있는 일종의 무증상이기에 상재균 취급을 받는데 이 마이코플라즈마균이 자가면역질환 등 다양한 질병을 유발한다고 주장하는 의사들도 있다.

The 12 types are treated as common bacteria because they are a type of asymptomatic bacteria that people may not know the symptoms of even if they are infected. Some doctors claim that these mycoplasma bacteria cause various diseases, including autoimmune diseases.

(논문) 마이코플라즈마는 생체 내에서 악성 형질 전환을 촉진하며, 박테리아 샤페론 단백질인 DnaK는 광범위한 발암 특성을 가지고 있다. (PMC6304983)

(Paper) Mycoplasma promotes malignant transformation in vivo, and DnaK, a bacterial chaperone protein, has extensive carcinogenic properties. (PMC6304983)

논문은, 세포 내에서 세포 변형 및 항암제에 대한 내성/숨김/내성 메커니즘에 관여할 수 있다고 기술하고 있다.

The paper states that it may be involved in cellular transformation and resistance/hidden/resistance mechanisms to anticancer drugs within cells.

(논문) "소 우유, 정액 및 면봉 샘플에서 Mycoplasma 종을 식별하기 위한 배양 및 다중 프로브 PCR 비교 (PMC5338856)"

(Paper) "Comparison of culture and multiple probe PCR for identification of Mycoplasma species in bovine milk, semen, and swab samples (PMC5338856)"

- 출처(Source): https://www.ncbi.nlm.nih.gov/

논문을 통해 수직감염이 될 수 있음을 예측할 수 있다.

Through the paper, it can be predicted that vertical infection may occur.

마이코플라즈마는 세포 표면에 붙어 기생하고, 감염되어 있더라도 세포의 형태학적인 변화가 거의 없고 일반적인 도립현미경(inverted microscope)에서는 관찰되지 않는다.

Mycoplasma attaches to the cell surface and lives as a parasite, and even if infected, there is little morphological change in the cell and is not observed under a general inverted microscope.

또한 세포 내 기생세균의 특성상 병원체의 완전제거가 어렵다.

Additionally, due to the nature of intracellular parasitic bacteria, it is difficult to completely eliminate pathogens.

마이코플라즈마는 숙주세포의 막에 미세소관을 형성하여 숙주세포의 영양분을 소비하는 과정에서 숙주세포의 손상을 유발하며 숙주세포 내로 침투하여 기생함으로써 숙주세포를 파괴하기도 한다. (이 과정에서 손상된 숙주세포가 상처로 인해 민감해지면서 비접촉 알레르기 등을 유발하는 것으로 추정된다.)

Mycoplasma forms microtubules in the host cell membrane and causes damage to the host cell in the process of consuming the host cell's nutrients. It also destroys the host cell by infiltrating the host cell and parasitizing it. (It is presumed that during this process, damaged host cells become sensitive due to the wound, causing non-contact allergies.)

또한 백일해독소와 유사한 구조를 지닌 community acquired respiratory distress syndrome 독소를 발현하여 숙주세포 파괴와 염증을 유발하기도 한다. (이 과정에서 세포의 고통인 가려움증과 염증현상이 발생되는 것으로 예측할 수 있다.)

In addition, community acquired respiratory distress syndrome toxin, which has a structure similar to pertussis toxin, is expressed, causing host cell destruction and inflammation. (It can be predicted that itching and inflammation, which are pain at the cellular level, will occur during this process.)

(논문) "Mycoplasma Pneumoniae 감염의 폐 외 증상과 어린이의 아토피/호흡기 알레르기 사이에는 어떤 관계가 있습니까? (PMC4821218)"와 (논문) "아토피가 소아 Mycoplasma Pneumoniae 폐렴의 중증도 및 폐렴 이외의 증상에 미치는 영향

(PMC6595288)"에 의하면, Mycoplasma Pneumoniae균은 숙주 세포의 세포막을 관통하고 호흡기 점막을 침범할 수 있다.

(Paper) "What is the relationship between extrapulmonary symptoms of Mycoplasma Pneumoniae infection and atopy/respiratory allergy in children? According to "(PMC4821218)" and (paper) "Effect of atopy on severity of Mycoplasma Pneumoniae pneumonia in children and symptoms other than pneumonia (PMC6595288)," Mycoplasma Pneumoniae bacteria can penetrate the cell membrane of host cells and invade the respiratory mucosa.

더욱이, MP감염은 현저한 염증반응을 야기할 수 있고 또한 어느 정도 호흡기 외부로 퍼질 수 있다. 따라서, MP감염은 호흡기 질환뿐만 아니라 호흡기 증상 이외의 증상을 초래할 수 있다.

Moreover, MP infection can cause a significant inflammatory response and can also spread outside the respiratory tract to some extent. Therefore, MP infection can cause not only respiratory disease but also non-respiratory symptoms.

어린이의 MP감염은 알레르기 질환의 발병 위험 요소이며 알레르기를 유발한다. 여러 연구에서 호흡기 MP감염이 천식 악화와 관련이 있어왔다.

MP infection in children is a risk factor for the development of allergic diseases and causes allergies. Respiratory MP infections have been associated with asthma exacerbation in several studies.

한편 신문기사에 의하면, (주)에이치앤비나인은 마이코플라즈마 감염과 피부질환 발병의 연관성을 연구해 오던 중 마이코플라즈마가 면역세포의 공격을 피해 섬유아세포의 안으로 침입하여 생존할 수 있으며, 세포와 세포 사이에 가늘고 긴 터널(Tunneling nanotube)을 생성시켜 이동한 후 증식한다는 것을 국내외 최초로 규명해냈다. (출처: https://www.mk.co.kr/news/society/9007686)

Meanwhile, according to a newspaper article, while HnB9 Co., Ltd. was researching the relationship between mycoplasma infection and the development of skin diseases, it was discovered that mycoplasma can survive by invading fibroblasts to avoid attack by immune cells, and that cells and cells It was discovered for the first time in Korea and abroad that a thin and long tunnel (Tunneling nanotube) is created between the nanotubes, moves, and then proliferates. (Source: https://www.mk.co.kr/news/society/9007686)

신문기사는 마이코플라즈마는 세균(Bacteria)의 일종으로 일부는 병원성이 좋아 폐렴, 피부건선, 아토피, 류마티스 관절염, 다발성경화증, 피부암과 같은 질환을 유발시킨다고 기재하고 있다.

The newspaper article states that Mycoplasma is a type of bacteria, and some of them are pathogenic and cause diseases such as pneumonia, skin psoriasis, atopy, rheumatoid arthritis, multiple sclerosis, and skin cancer.

## 인간 이외의 동물에 발생하는 아토피 유사현상 또는 마이코플라즈마균 감염증
## Atopy-like phenomenon or mycoplasma infection that occurs in animals other than humans

아토피 염증을 일으키는 존재의 다섯 번째 조건은 인간 이외의 동물에서 유사한 사례가 존재하는지 여부이다.

The fifth condition for the existence of atopic inflammation is whether similar cases exist in animals other than humans.

이 마이코플라즈마균이 100여 종이 있다면, 인간에게만 질병을 일으키지는 않을 것이고, 균이 많아졌다면 다른 동물도 유사한 유형의 질병이 유행해야 하기 때문이다.

If there are more than 100 types of mycoplasma bacteria, it will not only cause disease in humans, and if the number of bacteria increases, similar types of diseases should be prevalent in other animals as well.

살펴본 결과, 개, 돼지, 닭, 소 등 가축과 반려동물의 피부병변을 통해 아토피증후군과 유사한 증세를 동반한 마이코플라즈마 감염증이 관찰되고 있다.

As a result, mycoplasma infection with symptoms similar to

atopic syndrome is observed through skin lesions in livestock and companion animals such as dogs, pigs, chickens, and cows.

예를 들면 소의 경우 초기에는 다소 침울, 식욕감퇴 등의 대체적으로 가벼운 증상을 보이나, 이 질병이 급성으로 경과되면서 발열, 비강에서 삼출물의 배출, 과호흡, 기관지 호흡, 개구호흡 등과 같은 호흡곤란의 증상을 유발한다.

For example, in the case of cattle, generally mild symptoms such as depression and loss of appetite are shown in the early stages, but as the disease progresses acutely, symptoms of respiratory difficulty such as fever, discharge of exudate from the nasal cavity, hyperventilation, bronchial breathing, and open breathing, etc. causes

또 닭의 경우를 살펴보면, 호흡기 증상, 기침, 재채기, 호흡기음, 콧물이나 눈물, 낮은 생산성, 낮은 성장율, 다리 이상, 사료 섭취 감소, 부화율 감소와 병아리 생존율 감소, 부검 시 기낭염, 심질환, 간포질환(대장균 2차 감염 시), 비강, 검사의 건락성 복강, 때때로 교체과 건초질환, 수란관염이 나타난다.

Also, looking at the case of chickens, respiratory symptoms, coughing, sneezing, respiratory sounds, runny nose or tears, low productivity, low growth rate, leg abnormalities, decreased feed intake, decreased hatching rate and chick survival rate, air cystitis at necropsy, heart disease, and hepatocyst disease. (In case of secondary infection with E. coli), nasal cavity, caseous abdominal cavity of examination, sometimes replacement and tendon sheath disease, oviductitis appear.

개의 경우는 강아지 아토피가 광범위하게 퍼져 있음을 쉽게 인터넷 이미지 검색을 통하여 확인할 수 있다.

In the case of dogs, it can be easily confirmed through an Internet image search that canine atopic dermatitis is widespread.

돼지 또한 마이코플라즈마 폐렴에 걸리고 있다.

Pigs are also suffering from mycoplasma pneumonia.

이러한 결과는 쉽게 인터넷을 통하여 확인 가능하다.

These results can be easily confirmed through the Internet.

# 마이코플라즈마균 창궐의 원인
# Causes of Mycoplasma outbreaks on a large scale

아토피증후군은 전 지구적인 산업화와 함께 본격적으로 시작되었다.

Atopic syndrome began in earnest with global industrialization.

이 시점에서 마이코플라즈마균이 창궐한 이유를 알 필요가 있다.

At this point, it is necessary to know the reason for the Mycoplasma outbreak.

아토피증후군이 본격적으로 시작된 것이 지구온난화가 원인인지, 아니면 지구의 산소농도 변화가 원인인지 살펴보면, 지구의 산소농도는 12%에서 35% 사이를 주기적으로 변화하였다고 과학자들은 말하고 있다.

When examining whether the onset of atopic dermatitis was caused by global warming or changes in the Earth's oxygen concentration, scientists say that the Earth's oxygen concentration periodically varied between 12% and 35%.

그러나, 결론적으로 아토피라는 이름이 생긴 이후로 100년 역사 동안에 지구의 급격한 산소농도 변화는 없었다.

However, in conclusion, there has been no drastic change in oxygen concentration on Earth during the 100 years since the name atopy was coined.

산업사회 이전의 대기는 질소 78%, 산소 21%, 이산화탄소 0.028%(280ppm), 아르곤 등 기타물질 0.972%(9,720ppm)의 조성으로 구성되어 있었다.

The atmosphere before industrial society was composed of 78% nitrogen, 21% oxygen, 0.028% (280 ppm) carbon dioxide, and 0.972% (9,720 ppm) other substances such as argon.

그러나, 2022년 이산화탄소는 421ppm(50.35% 증가)이 되었다.

However, in 2022, carbon dioxide increased to 421 ppm (50.35% increase).

이 추세대로라면 2090년에는 750ppm(167.85%)이 될 것이라고 한다.

If this trend continues, it will be 750 ppm (167.85%) in 2090.

지구 온난화로 인한 온도상승이 아토피의 발병원인은 아니라고 보여진다.

It appears that temperature rise due to global warming is not the cause of atopy.

산업사회 이전과 산소 농도 또한 큰 차이가 없기 때문이다.

This is because there is no significant difference in oxygen concentration compared to before industrial society.

문제는 이산화탄소를 포함하는 온실가스의 증가로 자연정화 능력이 떨어지면서 마이코플라즈마균이 창궐하게 된 것으로 보여진다.

The problem appears to be that the natural purification ability has decreased due to the increase in greenhouse gases, including carbon dioxide, and mycoplasma bacteria have become rampant.

대기중의 오존($O_3$)은 매우 불안정하여 산소분자($O_2$)로 변화하기 위해 산소원자(O)를 내어 놓으려고 하는 경향이 있다.

Ozone ($O_3$) in the atmosphere is very unstable and tends to give up oxygen atoms (O) to change into oxygen molecules ($O_2$).

오존은 오존층에만 존재하는 것이 아니라 대기에 오존이 일부 포함되어 있다. 참고로 한국 2023년 3월 6일 기준 전국 대기의 오존농도는 평균 0.26ppm이다.

Ozone exists not only in the ozone layer, but some ozone

is also contained in the atmosphere. For reference, as of March 6, 2023 in Korea, the national atmospheric ozone concentration averaged 0.26 ppm.

이렇게 분리된 산소는 대기 중의 물 또는 수소와 만나서 과산화수소($H_2O_2$)가 되거나 물($H_2O$)이 되고, 발생된 과산화수소는 빗물과 함께 떨어져서 자연을 정화하는 역할을 하게 된다.

The separated oxygen meets water or hydrogen in the atmosphere and becomes hydrogen peroxide ($H_2O_2$) or water ($H_2O$), and the generated hydrogen peroxide falls with rainwater and plays a role in purifying nature.

이 과정에서 산업화로 인하여 늘어난 온실가스가 빗물과 함께 섞여 내려오는 자연 발생적인 과산화수소를 대지에 닿기 전에 흡수해서 자연정화 능력을 떨어뜨리는 것으로 추정된다.

In this process, it is presumed that greenhouse gases increased due to industrialization absorb naturally occurring hydrogen peroxide mixed with rainwater before it reaches the ground, reducing the natural purification ability.

이 현상이 땅에 사는 혐기성 미세기생충인 마이코플라즈마균을 정화하지 못하고 창궐하게 만든 원인으로 보여지며, 이로 인해 아토피증후군을 증가시킨 것으로 추정된다.

This phenomenon appears to be the cause of Mycoplasma bacteria, an anaerobic micro-parasite living in the ground, failing to be purified and causing an outbreak, which is presumed to have increased atopy syndrome.

자연 발생적인 과산화수소가 대지에 떨어지는 확률이 낮아지면서, 혐기성 세균인 마이코플라즈마 미세기생충이 대지에 창궐하게 되고, 감염된 야채, 과일, 뿌리채소 등을 통해 초식동물과 육식동물들도 감염시키고, 우리의 식탁으로 들어와 인간의 몸을 감염시키게 되는 것이다.

As the probability of naturally occurring hydrogen peroxide falling on the ground decreases, Mycoplasma microparasites, an anaerobic bacterium, become rampant on the ground, infecting herbivores and carnivores through infected vegetables, fruits, root vegetables, etc., and infecting our tables. It enters and infects the human body.

# 종래 NC/Nga mouse(아토피 피부염) 실험의 실패 이유
# Reasons for failure of conventional NC/Nga mouse (atopic dermatitis) experiment

아토피 원인균에 대한 연구 없이 수행한 수많은 NC/Nga mouse(아토피) 실험을 통한 신약 개발이 실패한 이유 또한 쥐에서 아토피 피부염을 일으키는 방법을 DNCB, DNFB 등의 약품으로 유발하였기 때문에 아토피 피부염과 유사한 피부 상태를 만들었을 뿐 실제 아토피와 달랐기 때문이다.

The reason why new drug development failed through numerous NC/Nga mouse (atopy) experiments conducted without research on the causative bacteria of atopic dermatitis is because the method of causing atopic dermatitis in mice was caused by drugs such as DNCB and DNFB, which is a skin condition similar to atopic dermatitis. This is because it was different from actual atopy.

쥐를 통한 아토피 실험은, 인간에게서 발견된 16종의 마이코플라즈마균을 쥐에게 이식하여 어떤 균이 어떤 병을 유발하는지에 대한 실험을 통해 마이코플라즈마균이 일으키는 다양한 질병을 관찰해야 하는 과제를 남겼다.

The atopy experiment using mice left the task of observing the various diseases caused by mycoplasma bacteria by transplanting 16 types of mycoplasma bacteria discovered in humans into mice and conducting experiments to determine which bacteria cause which disease.

# 마이코플라즈마균의 습성
## Habits of Mycoplasma

이 마이코플라즈마균의 습성은 여성의 유두, 생식기, 남성의 생식기, 목, 두피, 겨드랑이, 사타구니, 오금, 결막, 기관지, 비강 등 연약한 피부조식에 집중적으로 모이는 경향이 있다.

The habit of this mycoplasma bacteria tends to concentrate in soft skin tissues such as female nipples, genitals, male genitals, neck, scalp, armpits, groin, conjunctiva, bronchial tubes, and nasal cavity.

마이코플라즈마는 연약한 조직을 통해 지속적으로 피부의 표피 쪽으로 나오려고 하는 습성이 있는 것으로 관찰되고 있다.

It has been observed that mycoplasma has a habit of continuously trying to come out through soft tissues toward the epidermis of the skin.

이 습성은 폐조직(폐렴), 기관지(천식), 비강점막(비염), 연약한 피부세포(아토피, 건선, 피부암), 결막(결막염) 등 외부 공기와 접촉되는 신체 부위 또는 연약한 신체 부위에 발생하는 질병과 관련이 있는 것으로 관찰되었다. (피부암은 다른 원인일 수 있다.)

This habit is a disease that occurs in body parts or soft body parts that come in contact with external air, such as lung tissue (pneumonia), bronchial tubes (asthma), nasal mucosa (rhinitis), fragile skin cells (atopy, psoriasis, skin cancer), and conjunctiva (conjunctivitis). It was observed to be related to . (Skin cancer can be another cause.)

이 마이코플라즈마균 습성의 발견은 수년간 아토피 환자들을 관찰하는 과정을 통해 얻은 아토피 환자의 치료에 매우 중요한 정보를 제공하는 자료이다.

The discovery of the habits of this mycoplasma bacteria provides very important information for the treatment of atopic patients, obtained through the process of observing atopic patients for several years.

만일 마이코플라즈마균이 인체의 내부에 깊숙이 생존하거나 전신에 생존하는 습성이 있다면 치료는 더욱 어려울 것이다.

If mycoplasma bacteria have a habit of surviving deep inside the human body or throughout the body, treatment will be more difficult.

균을 끌어내어 살균할 수 있는 방법을 찾아야 한다. (아니면 균에 듣는 항생제를 찾아내거나…) 경우에 따라서는 두 가지가 다 필요할 수도 있다.

We need to find a way to draw out the germs and sterilize

them. (Or find an antibiotic that targets bacteria...) In some cases, both may be necessary.

지속적으로 밖으로 나오려는 습성은 반복된 피부치료를 통하여 결국 마이코플라즈마를 제균할 수 있다는 가능성을 말해주고 있는 것이기 때문이다.

This is because the habit of constantly coming out indicates the possibility that mycoplasma can eventually be eliminated through repeated skin treatments.

# 감염경로
## Infection route

어린아이가 태어날 때부터 아토피를 갖고 태어나는 것을 보고 유전을 의심하기도 했다.

After seeing a child born with atopy, I suspected it was hereditary.

그러나, 마이코플라즈마 미세기생충은 여성의 몸에서 상재균(resident flora) 대접을 받으며 멸균되지 않고 생존하는 능력을 보여주기도 한다.

However, Mycoplasma microparasites are treated as resident flora in the female body and show the ability to survive without being sterilized.

증상을 발현하지 않는 방법으로 살아남아 아직도 병원성이 확인되지 않는 마이코플라즈마균이 존재한다. 또한 남성의 정자 세포에도 기생할 수 있다.

There are Mycoplasma bacteria that survive by not showing symptoms and whose pathogenicity has not yet been confirmed. It can also parasitize male sperm cells.

즉 부모의 마이코플라즈마 감염은 자녀의 수직감염의 원인이 될 수 있다.

In other words, mycoplasma infection in parents can be the cause of vertical infection in children.

갓 태어난 아이의 아토피 피부염은 주로 볼에 나타나지만, 아이들은 전신이 근육이 거의 없는 연약한 피부조직을 가지고 있는 상태이기 때문에 빠른 시간 안에 전신 아토피 피부염 증세를 보인다.

Atopic dermatitis in newborn babies mainly appears on the cheeks, but since children have weak skin tissue with almost no muscles all over their body, they quickly show symptoms of atopic dermatitis throughout their body.

그러나, 아이가 자라는 과정에서 근육이 있는 피부가 생성되고 늘어나면 연약한 피부가 있는 사타구니, 오금 등으로 점차 균이 몰리는 경향이 있다.

However, as the child grows and the skin with muscles is created and stretched, bacteria tend to gradually flock to the groin, the back, etc., where the soft skin is located.

마이코플라즈마균은 인체의 몸 외부와 쉽게 접촉할 수 있는 연약한 곳을 찾아 기생하는 습성을 버리지 못한다.

Mycoplasma bacteria cannot give up their parasitic habit

of seeking out soft places where they can easily come into contact with the outside of the human body.

아토피증후군이 폐조직, 구강점막, 연약한 피부(유두, 팔 안쪽, 허벅지 안쪽, 겨드랑이, 오금, 얼굴, 사타구니), 결막, 기관지 등에 집중적으로 발생하는 이유다.

This is why atopic syndrome occurs concentrated in lung tissue, oral mucosa, soft skin (nipples, inner arms, inner thighs, armpits, crooks, face, groin), conjunctiva, and bronchial tubes.

마이코플라즈마균이 왜 이러한 연약한 부위를 통해 외부와의 접촉이 가능한 부분으로 몰려가는 습성을 가지게 되었는지는 밝혀진 바 없다. 다만, 식물에 기생할 때 표피에 기생했던 습성에 기인한 것은 아닌지 의심될 뿐이다.

It is not known why mycoplasma bacteria have the habit of flocking to areas where they can come in contact with the outside world through these soft parts. However, it is doubtful whether it is due to the habit of parasitizing on the epidermis when parasitizing plants.

최초의 감염과 성인 아토피의 감염은 먹거리에서 시작될 수 있다.

The first infection and adult atopic dermatitis can begin with food.

땅에서 기생하는 마이코플라즈마 미세기생충은 야채, 과일, 뿌리채소, 익히지 않은 땅에서 사는 동물의 고기, 동물의 피에 의해서 감염될 수 있기 때문이다.

This is because mycoplasma microscopic parasites that live in the ground can be infected by vegetables, fruits, root vegetables, uncooked meat of animals living in the ground, and animal blood.

신생아의 아토피는 부모 중 누구에게서 감염된 것인지는 알 수 없지만, 수직감염에 의한 것임을 의심할 수 있다.

Although it is unknown which of the parents infected the newborn with atopy, it is suspected that it is caused by vertical infection.

마이코플라즈마균은 상재균처럼 인체의 피부, 비강, 구강, 인두, 소화관, 오금, 비뇨 생식기 등에 존재할 수 있기 때문이다. 또한, 여성의 난자와 남자의 정자도 감염될 수 있다.

This is because mycoplasma bacteria, like common bacteria, can exist in the human skin, nasal cavity, oral cavity, pharynx, digestive tract, popliteal fossa, and urogenital tract. Additionally, female eggs and male sperm can also be infected.

발병률로 비교하면 아토피 피부염은 성인에 비해 대부분 소아에서 증상이 시작된다.

When compared by incidence, atopic dermatitis symptoms

mostly begin in children compared to adults.

이러한 소아 아토피의 증상은 아이가 자라면서 근육이 생기는 세포 부위와 연약한 피부를 가지는 부위로 나뉘는데 아토피의 원인균은 근육이 많은 부위를 싫어하고 연약한 피부를 가지는 부위로 점차 몰려 옮겨가서 연약한 피부에 몰려 환부를 형성하는 경향이 있다.

These symptoms of childhood atopy are divided into the cell area where muscles are formed and the area with soft skin as the child grows. The causative bacteria of atopy dislike areas with a lot of muscle and gradually migrate to areas with soft skin, flocking to the soft skin and causing the affected area. tends to form.

이렇게 외부 공기와의 접촉이 용이한 표피와 연약한 부위에 몰려서 아토피 증세가 나타나는 것은 마이코플라즈마균의 습성이 원인으로 보여진다.

The habit of mycoplasma bacteria appears to be the cause of the appearance of atopic dermatitis in the epidermis and vulnerable areas that are easily in contact with the outside air.

이것이 연령에 따라 발생 부위가 달라지는 원인으로 추정된다.

This is presumed to be the reason why the site of occurrence varies depending on age.

# 아토피 피부염의 유전 의심
## Suspected inheritance of atopic dermatitis

확실한 유전인자가 밝혀진 바 없지만, 끊임없이 아토피의 유전성이 거론되어 온 것이 사실이다.

Although no exact genetic factor has been identified, it is true that the hereditary nature of atopy has been constantly discussed.

그것은 아토피 피부염 환자의 70~80%에서 아토피(알레르기) 질환의 가족력이 있으며 부모 중 한쪽이 아토피 피부염인 경우 자녀에서 아토피 피부염의 발생이 증가하고, 부모 모두가 아토피 피부염인 경우 자녀의 80%에서 발생한다는 보고들이 있어 유전적 요인이 관여하는 것으로 여겨졌기 때문이다.

70-80% of atopic dermatitis patients have a family history of atopic (allergy) disease. If one parent has atopic dermatitis, the incidence of atopic dermatitis increases in children, and if both parents have atopic dermatitis, it occurs in 80% of children. This is because there have been reports that genetic factors are involved.

그러나, 아토피 피부염의 유전인자가 밝혀진 바 없는 것이 사실이고, 이러한 유전을 의심하게 하는 현상은 부모로부터의 수직감염에 의한 아토피 원인균의 감염으로도 충분히 설명될 수 있기 때문에 아토피 피부염은 유전보다는 수직감염에 무게를 두어야 한다.

However, it is true that the genetic factor for atopic dermatitis has not been identified, and the phenomenon that raises doubts about this inheritance can be fully explained by infection with atopy-causing bacteria due to vertical infection from parents, so atopic dermatitis is more likely to be caused by vertical infection rather than inheritance. weight must be given.

## 세포 내 기생충
## Intracellular parasites

세포 내 기생충(Intracellular parasite)은 숙주의 세포 내부에서 자라고 번식할 수 있는 미세기생충을 말하며, 우리가 생각하는 것보다 많은 종류의 기생충이 세포 내에 존재한다.

Intracellular parasites are microscopic parasites that can grow and reproduce inside the host's cells, and there are more types of parasites inside cells than we think.

이러한 세포 내 기생충은 기능적 세포 내 기생충과 의무적 세포 내 기생충으로 분류하는데 기능적 세포 내 기생충에는 대분류로 볼 때, 마이코박테리움을 포함하는 박테리아 12종류, 곰팡이류 3종류가 있고, 의무적 세포 내 기생충으로는 바이러스류와 폐포자충이라는 특정 곰팡이가 있다.

These intracellular parasites are classified into functional intracellular parasites and obligate intracellular parasites. In general, functional intracellular parasites include 12 types of bacteria, including Mycobacterium, and 3 types of fungi. Obligate intracellular parasites include There are viruses and a specific fungus called Pneumocystis carinii.

기능적 세포 내 기생충은 세포 내부 또는 외부에서 살고 번식할 수 있으며, 의무적 세포 내 기생충은 숙주세포 외부에서 번식할 수 없으며, 전적으로 세포 내 자원에 의존해서 번식한다.

Facultative intracellular parasites can live and reproduce either inside or outside the cell, while obligate intracellular parasites cannot reproduce outside the host cell and rely entirely on intracellular resources to reproduce.

세포 내 기생충 분류에 의하면 마이코플라즈마(Mycoplasma)는 세포 내부에 기생하면서 외부에서도 살고 번식할 수 있는 것으로 보이므로 기능적 세포 내 기생충으로 분류가 가능할 것으로 보인다.

According to the classification of intracellular parasites, Mycoplasma appears to be able to live and reproduce outside the cell while parasitizing inside the cell, so it appears that it can be classified as a functional intracellular parasite.

연약한 피부, 호흡기의 점막, 기관지(폐), 결막 등은 인간이 외부로부터 쉽게 오염될 수 있는 인체조직이라고 오해하기 쉽다.

It is easy to misunderstand that delicate skin, respiratory mucosa, bronchi (lungs), conjunctiva, etc. are human tissues that can be easily contaminated from the outside.

그러나, 이미 아토피증후군 질환이 발생한 인체에서 해당 조직은 원인균이 표피 쪽으로 나오려고 하는 습성이 있어서 해당 부위에 원인균이 집중되는 현상이 관찰된 것으로 설명하는 것이 옳다고 본다.

However, in the human body where atopic syndrome disease has already occurred, the tissue in question has a habit of causing causative bacteria to come out toward the epidermis, so it seems correct to explain that the phenomenon of causative bacteria concentrating in that area was observed.

즉, 마이코플라즈마균은 지속적으로 겉(표면)으로 나오려는 습성을 가지고 있는데 감염된 생명체 속에서도 연약한 부위를 통해 표피 쪽으로 이동하려는 경향을 지속적으로 보인다는 것이다.

In other words, mycoplasma bacteria have a habit of continuously coming out to the surface, and even in infected organisms, they continue to show a tendency to move toward the epidermis through soft parts.

이러한 마이코플라즈마균의 습성은 결국 아토피를 치료할 수 있다는 가능성을 보여주고 있다.

These habits of mycoplasma bacteria ultimately show the possibility of treating atopy.

지속적으로 밖으로 나오려는 습성은 반복적인 치료행위에 의하여 결국 마이코플라즈마를 살균할 수 있다는 가능성을 보여주고 있는 것이기 때문이다.

This is because the habit of continuously coming out

shows the possibility that mycoplasma can eventually be sterilized through repeated treatment.

## 지금까지 아토피의 치료가 어려웠던 이유
## The reason why atopy has been difficult to treat until now

아토피증후군의 환부를 살펴보면, 세포 속 기생충의 특성인 숙주세포의 영양분을 빼앗는 과정에서 세포 손상을 일으키는 특성과, 마이코플라즈마균의 습성인 연약한 피부이자 외부와 접촉 가능한 표피 쪽으로 이동하여 기생하려는 습성 두 가지를 모두 보여주고 있다.

When looking at the affected area of atopic syndrome, there are two characteristics: the characteristic of causing cell damage in the process of depriving the host cell of nutrients, which is the characteristic of intracellular parasites, and the habit of mycoplasma bacteria, which is the habit of moving to the epidermis, which is fragile and can come into contact with the outside and become parasitic. Everything is showing.

마이코플라즈마가 아토피의 원인이라 하더라도, 마이코플라즈마는 세포 속에 기생하는 미세기생충으로 세포벽이 없어 세포벽을 파괴하는 항생제 계통에 내성을 지니게 된다. 따라서 항생제로 치료하기 어렵고, 곧 내성이 생긴다.

Even though mycoplasma is the cause of atopy, mycoplasma is a microscopic parasite that lives inside cells and has no cell wall, making it resistant to antibiotics that destroy cell walls. Therefore, it is difficult to treat with antibiotics, and resistance soon develops.

각종 항생제뿐만 아니라, 알벤다졸, 이버맥틴 등 구충제에도 위축만 될 뿐 구충되지 않는다.

Not only various antibiotics, but also anthelmintic drugs such as albendazole and ivermectin only cause shrinkage and do not deworm the animal.

따라서 항생제나 구충제로 치료하는 것이 무척 어렵고 현미경 등으로 관찰하는 것도 용이하지 않다.

Therefore, it is very difficult to treat with antibiotics or anthelmintics, and it is not easy to observe it with a microscope.

또한 잘못된 실험방법의 설정으로 아토피 피부염 등에 대한 실험이 잘못되어 치료방향을 잘못 잡은 원인도 있다.

In addition, experiments on atopic dermatitis, etc. were performed incorrectly due to incorrect experimental method settings, which is also the reason why the treatment direction was not determined correctly.

이번 연구에 의하면, 직접 접촉 없이 인체의 세포가 알레르기

반응을 일으키는 것은, 마이코플라즈마균에 감염된 숙주세포가 영양분을 빼앗기는 과정에서 세포 손상이 발생하고, 이 세포 손상을 원인으로, 세포가 극도로 예민하게 반응하는 현상을 일으키는 것으로 추정되며, 이로 인하여 다양한 형태의 알레르기를 동반한 천식, 알레르기성 비염, 아토피 피부염, 알레르기성 결막염 등의 아토피증후군을 발생시키는 것으로 연구되었다.

According to this study, the reason why human cells cause allergic reactions without direct contact is because cell damage occurs when host cells infected with Mycoplasma bacteria are deprived of nutrients, and this cell damage causes the cells to become extremely sensitive. It is presumed to cause a reaction phenomenon, which has been studied to cause atopic syndromes such as asthma, allergic rhinitis, atopic dermatitis, and allergic conjunctivitis accompanied by various types of allergies.

아토피 환자의 음식물 알레르기 또한 실제 음식물 알레르기가 아니라 균이 숙주세포의 영양분을 빼앗는 과정에서 발생하는 가려움증으로 추정되었다.

Food allergy in atopic dermatitis patients was also assumed to be not an actual food allergy, but rather itching caused by bacteria depriving host cells of nutrients.

또한 마이코플라즈마균 특유의 습성으로 연약한 피부 주위로 모여 이동하려는 습성이 있기 때문에, 연약하면서 외부와 접촉이 가능한 부위 등에 집중적으로 모이는 특징을 보였다.

In addition, due to the unique habit of mycoplasma bacteria, they tend to gather and move around soft skin, so they tend to gather concentrated in areas that are soft and can come in contact with the outside.

하다못해 이미 병원성이 확인된 마이코플라즈마균들도 폐렴, 비임균성 요도염, 세균성 질염 등 연약한 인체 부위에 집중적으로 질병을 유발시킨다.

At the very least, mycoplasma bacteria whose pathogenicity has already been confirmed cause diseases concentrated in vulnerable parts of the human body, such as pneumonia, non-gonococcal urethritis, and bacterial vaginosis.

즉, 마이코플라즈마균 16가지 중에 특정한 균 한 가지의 감염 증상이 아토피증후군을 일으키는 것이 아니라, 인간에게 감염될 수 있는 마이코플라즈마균 16가지 대부분의 습성이 연약한 피부에 몰려 기생하는 습성과 함께, 세포 내 기생충의 특성인 숙주세포의 영양분을 빼앗는 특성을 보이고 있고, 이러한 특성이 숙주세포의 세포 손상을 일으켜, 손상된 세포가 예민해져 아토피증후군을 발생시키는 것으로 이해된다.

In other words, it is not the infection symptoms of one specific type of Mycoplasma bacteria that cause atopic syndrome among the 16 types of Mycoplasma bacteria, but the habit of most of the 16 Mycoplasma bacteria that can infect humans is the habit of parasitizing on delicate skin and intracellular parasites. It shows the characteristic of taking away nutrients from host cells, and it is understood that this characteristic causes cellular damage to the host cells, making the damaged cells sensitive and causing atopic syndrome.

아토피증후군(아토피 피부염)이 전신 가려움증이 아닌 특정 부위만 가려운 것은 이러한 마이코플라즈마균의 습성에 의한 것으로 관찰되고 이해된다.

It is observed and understood that the fact that atopic syndrome (atopic dermatitis) causes itching only in specific areas rather than the entire body is due to the habits of mycoplasma bacteria.

# 아토피증후군 및 세포 내 기생충에 의한 피부질환 치료방법에 대한 연구
# Research on treatment methods for atopic syndrome and skin diseases caused by intracellular parasites

앞서 설명한 바와 같이 아토피증후군을 일으키는 마이코플라즈마균은 세포 속에 기생하여 숙주세포를 만들기 때문에 치료 시에 다음과 같은 문제가 있다.

As explained earlier, mycoplasma bacteria that cause atopic syndrome live in cells and create host cells, so there are the following problems during treatment.

일반적인 광학현미경 등으로 관찰이 용이하지 않다.

It is not easy to observe using a general optical microscope.

연구가 덜 되어 병원성이 밝혀지지 않은 균이 많다.

There are many bacteria whose pathogenic properties are not yet known due to lack of research.

세포 속에 있는 균을 죽일 수 있는 마땅한 방법을 찾기 어렵다. (세포 내 기생세균의 특성상 완전제거가 어렵다.)

It is difficult to find a suitable method to kill bacteria inside cells. (Due to the nature of intracellular parasitic bacteria, complete removal is difficult.)

숙주세포 속으로 들어가는 약물의 설계가 어렵다.
Designing drugs that enter host cells is difficult.

항생제나 구충제에 쉽게 내성이 생긴다.
Resistance easily develops to antibiotics or insecticides.

세포 내 기생세균은 세포 외부에서 배양이 어렵기 때문에 연구가 어렵다.
Intracellular parasitic bacteria are difficult to study because they are difficult to culture outside the cell.

그러나, 이러한 다양한 어려움 속에서도 세포 내 기생충이 있음을 확인하였고, 마이코플라즈마균이 세포 속에서 기생하는 세포 내 기생충의 특성을 확인하였기 때문에, 세포 속으로 통하는 문(통로)을 찾는 연구를 하였다.
However, despite these various difficulties, the existence of intracellular parasites was confirmed, and the characteristics of intracellular parasites in Mycoplasma bacteria were confirmed, so research was conducted to find a door (passage) leading into the cell.

세균이 들어갈 수 있다면 반드시 통하는 문(통로)이 있을 것이기 때문이다.

This is because if bacteria can enter, there must be a door (passage) through which they can enter.

# 이온채널과 막수송체
# Ion channels and membrane transporters

세포에는 이온채널인 칼륨, 나트륨 이온채널이 있고, 이러한 이온채널은 세포의 세포막에 존재하면서 세포의 안과 밖으로 이온을 통과시키는 막단백질임을 알게 되었다.

It was discovered that cells have ion channels, such as potassium and sodium ion channels, and that these ion channels are membrane proteins that exist in the cell membrane and allow ions to pass in and out of the cell.

이러한 양이온 채널을 통과할 수 있는 세포막수송체로 나트륨 이온과 규소이온으로 구성된 유기규소이온액(organosilicon ionic liquid; chelated silicon)(한국 상품명: 유기규소이온 K9)을 선택하였다.

An organosilicon ionic liquid (chelated silicon) composed of sodium ions and silicon ions (Korean brand name: Organosilicon Ion K9) was selected as a narrow membrane transporter that can pass through this cation channel.

유기규소이온액(organosilicon ionic liquid; chelated silicon)(상품명: 유기규소이온 K9)은 한국 인터넷 쇼핑몰에서 구매할 수 있는 공지된 것이며, 먹을 수 있는 안전한 미네랄 영양소이자, 국제화장품 원료집에 등재된 제품(ICID Application NO. 4-07-2021-12380)으로서 화장품 원료로 사용할 수 있는 물질이다.

Organosilicon ionic liquid (chelated silicon) (Product name: Organosilicon Ion K9) is a known product that can be purchased at Korean online shopping malls, is a safe mineral nutrient for consumption, and is a product listed in the International Cosmetic Ingredient Directory (ICID). Application No. 4-07-2021-12380) is a substance that can be used as a cosmetic raw material.

규소이온은 인체 내에서 다양한 역할을 한다.

Silicon ions play various roles in the human body.

예를 들면, 칼슘을 뼈로 운반하는 화물차 역할(능동수송의 한 형태라 할 수 있다)을 하고, 콜라겐과 케라틴을 생성하는 중심물질에 해당한다.

For example, it acts as a truck that transports calcium to bones (it can be considered a form of active transport) and is the central substance that produces collagen and keratin.

즉 규소이온은 막수송체 역할을 한다.

In other words, silicon ions act as membrane transporters.

**지금까지 의학계는 막수송체는 단백질인 것으로 알고 있었다. 그러나, 킬레이트 규소이온은 단백질이 아니면서 막수송체 역할을 한다.**

**Until now, the medical community knew that membrane transporters were proteins. However, chelating silicon ions are not proteins but act as membrane transporters.**

규소이온이 막수송체로서 작용한다는 개념은 세포 생물학의 확립된 원리에서는 받아들이기 힘들 것이다.

The concept that silicon ions act as membrane transporters would be difficult to accept under established principles of cell biology.

그러나, 현재 개발된 실리콘 킬레이트 이온액에 세포 성장인자와 유황이온 등을 첨가하여 사용하는 것만으로 세포의 이온채널을 통과하여 발생할 수 있는 현상을 보여주고 있다.

However, just by adding cell growth factors and sulfur ions to the currently developed silicon chelate ionic liquid, it has been shown to pass through the cell's ion channel.

규소이온액이 개발된 시점에서는 실험을 통한 막수송체 역할을 검증하는 것은 도움이 되지만, 막수송체가 가능하다 불가능하다 토론은 큰 의미가 없다.

At the point when silicon ionic liquid was developed, it is helpful to verify the role of membrane transporters through experiments, but the discussion of whether membrane transporters are possible or not is of little significance.

이 규소이온이 세포 성장인자와 유황이온을 세포에 전달하는 현상은 실험이 안전하므로 규소이온액에 첨가하여 피부에 도포하는 방식의 실험을 통해 쉽게 효능을 입증할 수 있다.

This phenomenon of silicon ions delivering cell growth factors and sulfur ions to cells is safe to experiment, so its efficacy can be easily proven through experiments by adding it to silicon ion liquid and applying it to the skin.

유황이온은 살균소독제 역할을 하는데 규소이온이 막수송체로서 수송할 수 있는 미네랄 이온의 한 종류이다.

Sulfur ions act as a sterilizing disinfectant and are a type of mineral ion that silicon ions can transport as a membrane transporter.

또 규소이온으로 생성되는 콜라겐은 세포 간 접착제의 역할을 기대할 수 있다.

Additionally, collagen produced from silicon ions can be expected to act as an intercellular adhesive.

접착제 역할은 노인성 소양증과 세포 단위의 고통으로 인한 스트레스(예: 불면증) 해소에 도움이 된다.

Its role as an adhesive helps relieve senile pruritus and stress caused by pain at the cellular level (e.g. insomnia).

그러나, 이러한 규소이온액을 피부에 단독으로 발랐을 때, 아토피 피부염을 완화하거나 치료하는 현상이 눈에 띄게 발생하지는 않았다.

However, when this silicon ion solution was applied alone to the skin, there was no noticeable phenomenon of alleviating or treating atopic dermatitis.

규소이온은 막수송에 있어서 수동수송 또는 능동수송 수송체 역할을 기대할 수 있고, 세포 내부에서 영양분으로 작용하는 역할도 기대할 수 있다.

Silicon ions can be expected to play a role as a passive or active transporter in membrane transport and can also be expected to act as nutrients inside cells.

그러나, 규소이온이 직접적인 아토피 치료제 역할을 하지는 않는다.

However, silicon ions do not act as a direct atopy treatment.

다만, 이온채널을 열고 막수송체 역할을 충실하게 할 뿐이다.

However, it only opens ion channels and faithfully functions as a membrane transporter.

규소이온이 얼마나 많은 종류의 물질을 수송할 수 있는지는 추가 연구가 필요하다.

Further research is needed to determine how many types of materials silicon ions can transport.

# 세포의 치료를 위한 성분 1 – 성장인자
## Ingredient 1 for cell treatment – cell growth factors

성장인자(Growth Factor, GF)는 세포증식, 상처치유, 세포분화를 자극할 수 있는 자연발생 물질이다.

Growth Factor (GF) is a naturally occurring substance that can stimulate cell proliferation, wound healing, and cell differentiation.

각각의 성장인자는 세포분화와 성숙을 촉진한다.

Each growth factor promotes cell differentiation and maturation.

표피세포 성장인자(EGF), 섬유아세포 성장인자(FGF), 혈관내피 성장인자(VEGF), 인슐린 유사 성장인자(IGF) 등이 있다.

These include epidermal growth factor (EGF), fibroblast growth factor (FGF), vascular endothelial growth factor (VEGF), and insulin-like growth factor (IGF).

본 실험에서 EGF 표피세포 성장인자는 화장품 성분으로 사용되는 시중에서 쉽게 구할 수 있는 화장품 원료로 유통되는 10ppm 미만의 EGF를 사용하였다.

In this experiment, EGF epidermal cell growth factor was used at less than 10ppm EGF, which is an easily available cosmetic ingredient on the market.

본 실험에서 FGF는 섬유아세포 성장인자 1~23을 따로 구분하지 않은 화장품 성분으로 사용되는 시중에서 쉽게 구할 수 있는 화장품 원료인 10ppm 미만의 FGF를 사용하였다.

In this experiment, less than 10 ppm of FGF, which is a commercially available cosmetic ingredient used as a cosmetic ingredient without separate fibroblast growth factors 1 to 23, was used.

본 실험에서 IGF는 인슐린 유사 성장인자 1~2를 따로 구분하지 않은 화장품 성분으로 사용되는 시중에서 쉽게 구할 수 있는 화장품 원료인 10ppm 미만의 IGF를 사용하였다.

In this experiment, less than 10 ppm of IGF, which is a commercially available cosmetic ingredient used as a cosmetic ingredient without separate insulin-like growth factors 1 and 2, was used.

그러나, 종래 기술에서 보듯이 성장인자를 피부 표피에 바르는 형태로 단독으로 사용해서는 아토피 피부염에 획기적인 치료효과를 기대하기는 어렵다.

However, as seen in the prior art, it is difficult to expect a groundbreaking treatment effect for atopic dermatitis when growth factors are used alone in the form of applying them to the skin epidermis.

본인 또한 EGF, IGF, FGF 등을 구매하여 개별적으로 확인해보았으나, 아토피 피부염 등에 눈에 띄는 효과를 확인할 수 없었다.

I also purchased EGF, IGF, FGF, etc. and checked them individually, but I could not confirm any noticeable effect on atopic dermatitis, etc.

이렇게 아토피 피부염의 완화 내지 치료효과가 두드러지지 않는 것은, 세포 외부에서 사용했기 때문에 세포 내부에서 작용하지 못하는 원인이 가장 크다는 결론이다.

The conclusion is that the reason why there is no noticeable relief or treatment effect for atopic dermatitis is because it is used outside the cells and does not work inside the cells.

본 실험에서 유기규소이온액과 성장인자를 희석하여 피부에 사용할 경우 세포 내 기생충과 반응하여 1분 내지 3분 이내에 가려움증이 해소되는 현상과 함께 지속적으로 사용할 경우 1주일 이내에 상처가 아물듯이 딱지가 생기는 현상을 확인할 수 있다.

In this experiment, when diluted organosilicon ion solution and growth factors are used on the skin, it reacts with intracellular parasites and relieves itching within 1 to 3 minutes. If used continuously, scabs form within a week just as a wound heals. The phenomenon can be confirmed.

# 세포의 치료를 위한 성분 2 – 유황(Sulfur) 성분을 함유한 소독제

## Ingredient 2 for cell treatment – Disinfectant containing sulfur

유기규소이온액이 막수송체 역할을 통해 세포 성장인자를 이용하여 세포 내부에 작용하여 치료효과를 얻어내게 된 것은 규소이온이 인체 내에서 칼슘이온을 뼈로 운반하는 막수송체 역할을 하는 것과 같은 원리에 기인한 것이다.

The reason why organosilicon ionic liquid acts on the inside of cells using cell growth factors through its role as a membrane transporter to achieve therapeutic effects is the same principle as silicon ions acting as a membrane transporter that transports calcium ions to bones in the human body.

세포 속에 기생하는 기생충은, 알벤다졸이나, 이버맥틴 같은 구충제 또는 항생제는 쉽게 내성이 생겨 치료효과를 얻기 어렵다.

Parasites living inside cells easily become resistant to anthelmintics or antibiotics such as albendazole or ivermectin, making it difficult to achieve therapeutic effects.

그러나, 막수송체(규소이온)가 운반할 수 있는 소독제라면 그 치료효과를 기대할 수 있다.

However, if it is a disinfectant that can be transported by a membrane transporter (silicon ion), its therapeutic effect can be expected.

유황은 인체를 구성하는 성분이면서 다른 한편으로는 살균제 역할을 한다.

Sulfur is a component of the human body, but on the other hand, it acts as a disinfectant.

본 발명의 살균제 성분은 규소의 막수송체 역할을 통해 세포 내로 수송되어 세포 내 기생 미생물을 살균할 수 있는 성분이어야 한다.

The disinfectant component of the present invention must be an ingredient that can sterilize intracellular parasitic microorganisms by being transported into cells through the role of a silicon membrane transporter.

그런 의미에서 황(Sulfur) 성분을 포함하는 살균제는 자체적으로도 세포 내 기생 미생물을 살균하여 치료하는 효과가 있고, 세포 성장인자를 더할 경우 더 빠른 치료효과를 기대할 수 있다.

In that sense, disinfectants containing sulfur are effective in treating intracellular parasitic microorganisms by sterilizing them, and if cell growth factors are added, a faster treatment effect can be expected.

그러므로 규소가 막수송체 역할에 부합하여 이송할 수 있는 성분을 선택하여야 한다.

Therefore, a component that can be transported by silicon that matches its role as a membrane transporter must be selected.

예를 들면, 황 성분의 살균제로서 유황이온, 티오황산나트륨 등을 들 수 있다.

For example, sulfur-containing disinfectants include sulfur ions and sodium thiosulfate.

황(Sulfur) 성분을 포함하는 살균제의 사용은 성장인자의 사용으로 피부질환 치료용 조성물을 완성할 수 있으나, 중증 아토피 등 심한 아토피 질환의 경우 치료기간이 길어져 환자가 그 치료기간을 견디기 힘들어하는 경우가 생기기 때문에 치료기간을 단축하고자 살균제로서 사용하기 위함이다.

The use of a disinfectant containing sulfur can complete the composition for treating skin diseases by using growth factors. However, in the case of severe atopic diseases such as severe atopy, the treatment period is long and the patient has difficulty enduring the treatment period. It is intended to be used as a disinfectant to shorten the treatment period.

그러나, MSM(Methylsulfonylmethane)은 사용할 수가 없다.

However, MSM (Methylsulfonylmethane) cannot be used.

MSM은 식이유황으로 메틸기($CH_3$)가 2개 붙어 있어 물에 쉽게 녹는 것처럼 보이지만, 해리되지 않고 물 속에서 안정하다.

MSM is dietary sulfur and has two methyl groups ($CH_3$) attached to it, so it appears to dissolve easily in water, but it does not dissociate and is stable in water.

이러한 현상은 규소를 유기규소로 만드는 과정에서 메틸기($CH_3$)를 사용하였던 MMST, Silanol 등에서도 나타나는 현상으로 유기규소는 맞으나 이온 상태가 되지 못한다.

This phenomenon also occurs in MMST and Silanol, which use methyl groups ($CH_3$) in the process of converting silicon into organosilicon. Although they are organosilicon, they are not in an ionic state.

따라서 MSM은 유기유황이기는 하지만 이온이 아니기 때문에 막수송체 역할을 하는 규소이온에 의해 세포 내부로 전달되기 어렵다.

Therefore, although MSM is an organic sulfur, it is not an ion, so it is difficult to be delivered into cells by silicon ions that act as membrane transporters.

그러므로 식이유황은 규소이온의 막수송체 역할에 의하여 운반되지 않으므로 세포 내부에서 작용하기 어렵다.

Therefore, dietary sulfur is not transported by the membrane transporter of silicon ions, so it is difficult to act inside cells.

메틸설포닐메탄(MSM)은 유기황 화합물이지만 이온은 아니다. 표준 형태에서 MSM은 전하를 운반하지 않으므로 비이온성 화합물이다. 결과적으로 이온의 통과를 특별히 촉진하는 막수송체와 쉽게 상호작용하지 않는다.

Methylsulfonylmethane (MSM) is an organic sulfur compound, but it is not an ion. In its standard form, MSM does not carry an electrical charge, making it a non-ionic compound. As a result, it does not readily interact with membrane transporters that specifically facilitate the passage of ions.

수산화나트륨의 발열반응을 이용하여 유황이온을 제조하여 사용할 수도 있다. 이러한 황(Sulfur)은 120℃ 이상의 온도에서 알칼리 상태에서 용해가 된다.

Sulfur ions can also be produced and used using the exothermic reaction of sodium hydroxide. This sulfur dissolves in an alkaline state at a temperature of 120° C or higher.

따라서 황(Sulfur)의 이온 상태를 유지하려면 pH를 알칼리로 유지하여야 한다.

Therefore, to maintain the ionic state of sulfur, the pH must be maintained at alkaline.

2리터의 비이커에 수산화나트륨(NaOH) 200g과 황(S) 250g을 넣고 물($H_2O$) 500ml를 넣고 나무주걱이나 유리막대로 저어준다.

Put 200g of sodium hydroxide (NaOH) and 250g of sulfur (S) in a 2-liter beaker, add 500ml of water ($H_2O$), and stir with a wooden spatula or glass rod.

발열반응에 의하여 120℃에서 황(sulfur)이 녹기 시작하여 완전용해가 일어나면 물($H_2O$) 380ml를 추가로 넣어주고 식힌다.

Sulfur begins to melt at 120° C due to an exothermic reaction, and when complete dissolution occurs, add an additional 380ml of water ($H_2O$) and cool.

따라서, 규소이온이 세포의 이온채널을 여는 역할을 하고, 이온채널이 열린 상태에서 규소이온에 의해 수동운송 또는 능동운송에 의한 수송체 역할을 통하여, 황(Sulfur) 성분을 포함하는 살균제와 세포 성장인자가 세포의 내부에 수송되어 영향을 끼치게 하는 방법을 기대하며 피부질환 치료용 조성물을 구성할 수 있다.

Therefore, silicon ions play the role of opening the ion channel of the cell, and when the ion channel is open, silicon ions act as a transporter through passive or active transport, allowing disinfectants containing sulfur and cell growth. A composition for treating skin diseases can be constructed by anticipating a method of allowing factors to be transported and affect the inside of cells.

이를 위하여 세포로 통하는 이온채널을 여는 역할을 담당할 유기규소이온액(organosilicon ionic liquid; chelated silicon)에 황(Sulfur) 성분의 살균제, 세포 성장인자를 첨가하여 아토피 피부염을 포함한 피부질환자에게 사용할 수 있는 피부질환 치료용 조성물을 제조하여 완성하였다.

To this end, a sulfur-containing disinfectant and cell growth factors are added to organosilicon ionic liquid (chelated silicon), which plays the role of opening ion channels leading to cells, to create a product that can be used for patients with skin diseases including atopic dermatitis. A composition for treating skin diseases was prepared and completed.

유황이온의 사용은 치료기간을 단축시켜 주지만, 세균의 사체가 만들어내는 엔도톡신 반응으로 인하여 농도에 따라 심한 통증을 유발한다.

The use of sulfur ions shortens the treatment period, but causes severe pain depending on the concentration due to the endotoxin reaction produced by bacterial corpses.

## 실시예 1 피부질환 치료용 조성물 제조

## Example 1 Preparation of composition for treating skin diseases

**구성**: 유기규소이온액 500g, EGF 1ppm 용액 7.5g, FGF 1ppm 용액 7.5g, IGF 1ppm 용액 7.5g을 유효성분으로 포함하고, 부가 첨가물로 라루릴글루코사이드(유화제) 0.75g, 비타민 B3 2.5g, 글루타치온 0.75g을 더 첨가 희석하여 피부질환 치료용 조성물을 제조하였다.

**Composition**: Contains 500g of organosilicon ion solution, 7.5g of EGF 1 ppm solution, 7.5g of FGF 1 ppm solution, and 7.5g of IGF 1 ppm solution as active ingredients, and additional additives include 0.75g of lauryl glucoside (emulsifier), 2.5g of vitamin B3, and 0.75g of glutathione. g was added and diluted to prepare a composition for treating skin diseases.

이 조성물은 화장료로 사용할 수 있는 농도의 조성물에 해당한다.

This composition corresponds to a composition with a concentration that can be used as a cosmetic.

pH는 3.5를 나타내었는데 pH 상승과 완화는 수산화나트륨(NaOH) 또는 L-arginine을 사용하여 조정할 수 있다.

The pH was 3.5, and pH elevation and relief can be adjusted using sodium hydroxide (NaOH) or L-arginine.

pH 3.5는 피부에 도포하였을 때, 약간 따끔거리는 증세가 있으나, 소독의 효과가 있고, pH 5.5 내지 7.0은 저자극이어서 어린 아이들이나 민감한 피부를 가진 사람들이 사용하기에 적합하였다.

pH 3.5 causes a slight stinging sensation when applied to the skin, but has a disinfecting effect, and pH 5.5 to 7.0 is hypoallergenic, making it suitable for use by children and people with sensitive skin.

pH 조절을 통하여, 결막염에 사용할 수 있는 점안액 제조를 위하여 5.5 내지 8.0, 화장품으로 사용할 수 있는 화장료 제조를 위하여 5.5, 비염에 사용할 수 있는 분무제 제조를 위하여 5.5 등으로 조절하여 다양한 형태로 변형할 수 있고, 부가 첨가물들을 달리할 수 있다.

By adjusting the pH, it can be modified into various forms by adjusting it to 5.5 to 8.0 to manufacture eye drops that can be used for conjunctivitis, 5.5 to manufacture cosmetics that can be used as cosmetics, and 5.5 to manufacture sprays that can be used for rhinitis. , additional additives can be varied.

## 적용례 1 성인 아토피 피부염

## Application example 1 Preparation of composition for treating skin diseases

**적용대상**: 61세 여성 성인 아토피

**Applicable target**: 61-year-old female adult with atopic dermatitis

**적용방법**: 매일 1회 가려움증 환부에 기초화장품 바르듯이 바르고 중간에 가려움증이 발생되면 한 번씩 더 도포한다.

**How to apply**: Apply as if applying basic cosmetics to the itchy area once a day and apply once more if itching occurs.

**적용결과**: 바르는 즉시 가려움증이 해소됨. 가려움증은 8시간 정도 이후 다시 나타남. 이후 몇 번 가려움증이 반복되나 가려움증 범위가 점점 줄어드는 현상이 관찰됨. 홍조와 아토피 연고로 탄력을 잃었던 피부가 다시 팽팽하게 탄력이 생김. 거친 표면도 매끄럽게 변화되었음.

**Result of application**: Itching is relieved immediately upon application. The itching reappears after about 8 hours. Afterwards, the itching repeats several times, but the extent of the itching gradually decreases. Skin that had lost elasticity due to redness and atopic ointment regains its elasticity. Even rough surfaces became smooth.

특이한 점은 각질 탈각이 빠른 속도로 이루어지고 탈각 후 맑은 피부로 다시 돌아옴(피부가 건강한 피부세포로 빠르게 재생이 이루어지는 것 같음).

What is unique is that dead skin cells are shed at a rapid rate and clear skin returns after the shedding (skin appears to be rapidly regenerating into healthy skin cells).

## 적용례 2 성인 아토피 피부염

## Application example 2 Adult atopic dermatitis

**적용대상:** 남성 77세 가려움증이 심한 아토피

**Applicable to:** 77-year-old male with severe itching and atopic dermatitis

**적용방법:** 매일 1회를 바르고 3분 있다가 1회를 더 도포함.

**How to apply:** Apply once daily, wait 3 minutes, then apply once more.

**적용결과:** 가려움증이 있는 환부에 바르면 1분 이내에 가려움증이 확실하게 사라짐(전혀 가렵지 않음). 이 상태에서 3분쯤 지나 환부에 1회 더 도포함. 이렇게 2번 바르면 6시간 이상 가려움증이 생기지 않음(환자마다 약효의 시간이 조금씩 차이가 있음).

**Application results:** When applied to the itchy affected area, the itching disappears for sure within 1 minute (it does not itch at all). After about 3 minutes in this state, apply one more time to the affected area. If you apply it twice like this, itching will not occur for more than 6 hours (The time of drug effect varies slightly from patient to patient).

가려움증이 생겼던 부위에 3일 정도가 지나면, 아주 작은 딱지가 생기면 며칠간 가려움증이 멈춤. 약 6일 내지 7일 후 딱지(scab)가 떨어진 후 깨끗한 피부가 생김. 며칠 후 다시 가려움증이 시작됨. 이런 증상이 반복되면서 가려움증 부위가 점점 줄어듦.

After about 3 days, a tiny scab forms on the itchy area and the itching stops for a few days. After about 6 to 7 days, the scabs fall off and clear skin appears. After a few days, itching begins again. As these symptoms repeat, the itchy area gradually decreases.

6개월 사용한 시점에서의 경과사항은 하루에 1회만 발라도 가려움증이 하루 종일 발생하지 않아서 정상적인 생활이 가능함.

After using it for 6 months, itching does not occur all day even if you apply it once a day, so you can lead a normal life.

이 사례는 연령이 77세에 달해 몸 속에 균이 얼마나 많을지 몰라서 규소이온 섭취를 하지 않고 피부의 가려움증만 치료 가능한지 확인한 사례임(규소이온을 섭취했을 경우 얼마나 많은 원인균이 나올지 몰라서 완벽한 치료방법이 생기기 전까지 피부의 가려움증만 치료한 사례에 해당함).

In this case, the patient was 77 years old and did not know how many bacteria there were in his body, so he did not ingest silicon ions and checked whether only itching on the skin could be treated (This is a case where only skin itching was treated until a perfect treatment method was developed, as it was not known how many causative bacteria would be produced when silicon ions were ingested).

## 적용례 3 성인 아토피 피부염

## Application example 3 Adult atopic dermatitis

**적용대상**: 67세 여성

**Applicable target**: 67 year old women

**적용방법**: 매일 1회 가려움증 부위에 도포

**How to apply**: Apply to itchy area once daily

**적용결과**: 가려움증이 생기는 부위에 바르고 나면 즉시 또는 1분 이내에 가려움증이 해소됨. 또한 가려움증이 생기는 빈도가 점점 느려지고 있음.

**Results of application**: After applying to the itchy area, itching is relieved immediately or within 1 minute. Also, the frequency of itching is becoming slower.

## 적용례 4 성인 아토피 피부염 4개월 사용

## Application example 4 Adult atopic dermatitis: 4 months of use

**적용대상:** 59세 여성 하반신 전체 심한 가려움.

**Applicable to:** 59-year-old female. Severe itching of the entire lower body.

**적용방법:** 아침에 한 번, 저녁에 자기 전에 한 번 가려운 부위에 도포하는 방식으로 사용함.

**How to apply:** Apply to itchy areas once in the morning and once in the evening before going to bed.

**적용결과:** 바르기 전에는 가려움증이 너무 심해서 뜨거운 물로 목욕하는 방법으로 가려움증을 달래야 할 정도로 힘들었음.

**Application results:** Before applying, the itching was so severe that I had to relieve it by taking a hot water bath.

그러나, 가려움증이 있는 부위에 도포를 하면 바르는 순간 약간 따가웠다가 시원해지면서 1분 이내에 가려움증이 완전히 없어짐. 3일 동안 바르고 나면, 딱지(scab)가 생김. 딱지(scab)는 약 일주일 만에 떨어짐.

However, when applied to an itchy area, it stings slightly

the moment it is applied, but then cools down and the itchiness completely disappears within a minute. After applying it for 3 days, a scab forms. The scab falls off in about a week.

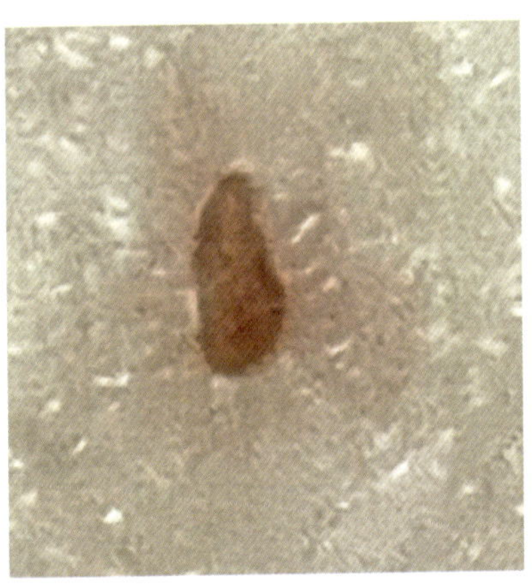

**도 1** **은 아토피 피부염이 치료과정에서 나타나는 딱지(scab)의 형태임.**
**Figure 1** **This is the form of scabs that appear during the treatment of atopic dermatitis.**

세포 단위로 선이 그어진 부분에 맞게 딱지(scab)가 생기고 떨어져 나가고 새살이 돋아남.

A scab forms in accordance with the area where the line is drawn on a cell-by-cell basis, falls off, and new skin sprouts.

다시 딱지가 떨어지고 깨끗한 피부가 나오면, 며칠 지나면 깨끗한 피부 밑에서 다시 가려움증이 느껴진다.

When the scab falls off again and clean skin appears, after a few days you will feel itching again under the clean skin.

다시 가려움증 부위에 도포하면 처음과 같이 가려움증은 1분 이내에 완전히 없어짐. 그리고 며칠 후 딱지(scab)가 생김(가려움증은 도포 후 8시간이 경과하면 다시 발생함. 발생하는 시간이 점점 느려짐).

When applied again to the itchy area, the itching disappears completely within 1 minute, just like the first time. And after a few days, a scab appears. (Itching occurs again 8 hours after application. The occurrence time gradually becomes slower)

이 현상이 무한정 반복될 것이라고 생각했는데 시간이 지나면서, 가려움증 부위가 처음에는 하반신 전체에 줄어들기 시작해서 국소부위로 점점 줄어듦.

I thought this phenomenon would repeat indefinitely, but as time passed, the itchy areas began to decrease, first over the entire lower body, and then gradually decreased to localized areas.

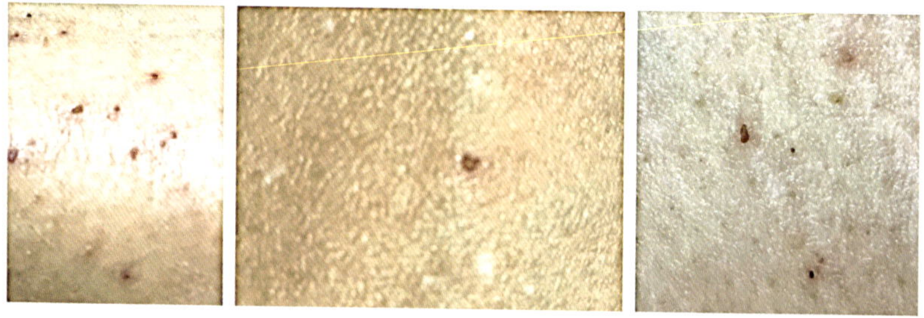

**도 2** 의 딱지는 4개월 사용 후 현재 피부에 남아있는 딱지의 형상.
**Figure 2** Shape of the scab remaining on the skin after 4 months of use.

초기에는 피부 대부분이 딱지로 뒤덮혔지만, 4개월 후에는 국소 부위 가려움증이 있을 때마다 발라주면, 피부가 두꺼워지거나, 태선화로 진행되지 않고, 가려움증이 즉시 멈추며 3~4일 이내에 딱지가 생기고 이후 일주일 이내에 새살이 돋아남.

Initially, most of the skin was covered with scabs, but after 4 months, if you apply it whenever there is itching in the local area, the skin does not thicken or progress to lichenification, the itching stops immediately, scabs form within 3 to 4 days, and after a week, the skin does not thicken or progress to lichenification. New skin sprouts within.

가려움증 부위도 완선히 줄어들어 국소 부위에 가끔 나타나고 전체적으로는 가려움증이 95% 이상 사라진 상태임.

The itchy area is also completely reduced, appearing occasionally in localized areas, and overall, more than 95% of the itching has disappeared.

아토피 피부염으로 인한 가려움증은 95% 이상 없어지고, 습진 증세, 태선화 현상은 100% 모두 사라짐.

Itching caused by atopic dermatitis disappears by more than 95%, and all symptoms of eczema and lichenification disappear by 100%.

국소 부위의 가려움증은 마치 무엇인가가 몰려와서 가려움증을 만드는 것으로 보이기도 하는데 <도 3>의 피부 속에 있는 형상이 먼저 나타나면서 가려움이 시작되는 경우임.

Itching in a localized area may appear as if something is rushing in to cause itching, but this is the case where the shape in the skin in Figure 3 appears first and the itching begins.

**도 3** **가려움이 생기기 전 나타난 피부 상태**
**Figure 3** **Skin condition before itching occurs**

〈도 3〉 피부 위에 도포를 시작하면 가려움증이 멈추면서 3일 정도 지나면 딱지가 생기고 점점 정상 피부로 회복됨.

Figure 3 When you start applying it on the skin, the itching stops, and after about 3 days, scabs form and the skin gradually returns to normal.

4개월 사용한 현재 상태는 전신에 부분적인 가려움증이 최근 1개월 내에 발생하지 않았고, 몇몇 딱지만 남은 상태이며, 더 이상 가려움증이 발생되지 않을 것 같음.

The current status after using it for 4 months is that partial itching on the whole body has not occurred within the past month, only a few scabs remain, and it seems that no more itching will occur.

또 가려움증이 다시 발생된다 하더라도 국소 부위에 생기고 그 부위에 치료제만 발라주면 쉽게 딱지가 앉게 된다는 확신을 가지게 되었음.

Also, I became confident that even if the itching occurred again, it would occur in a local area and a scab would easily form if I just applied the treatment to that area.

## 적용례 5 화폐상 습진

## Application example 5 Numismatic eczema

**적용대상**: 40대 남성 화폐상 습진

**Applicable to**: Men in their 40s with eczema

**적용방법**: 매일 1회 화폐상 습진 부위에 바름.

**How to apply**: Apply to eczema area once daily.

**적용결과**: 매일 화폐상 습진 부위에 바르면 바른 즉시 가려움증이 가라앉아 참을 수 있게 됨. 또한 피부도 빠르게 진정됨.

**Result of application**: If applied daily to the eczema area, the itching subsides immediately after application and becomes bearable. Also, the skin calms down quickly.

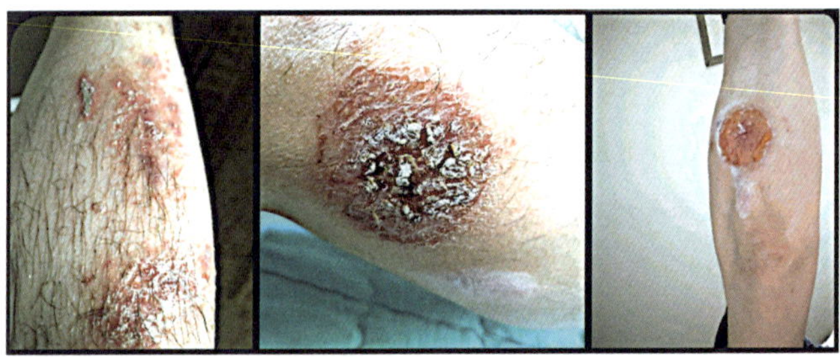

도 4 는 화폐상 습진이 발생된 피부 모습

Figure 4 shows the skin with nummular eczema

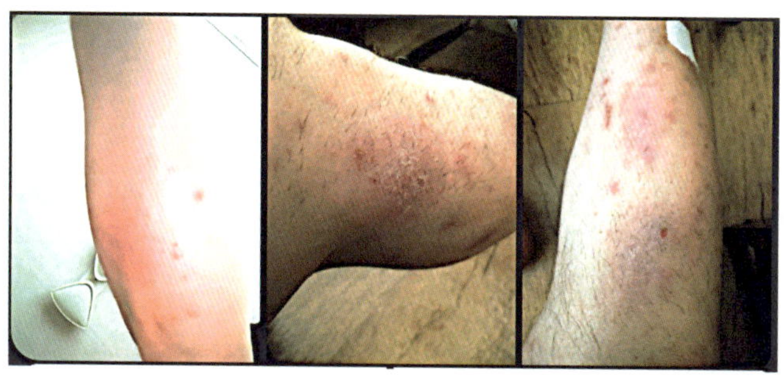

**도 5** 는 피부염 완화 조성물을 바르고 약 10일 경과 후 진정되고 있는 화폐상 습진 모습

**Figure 5** Nummular eczema subsiding about 10 days after applying the dermatitis alleviating composition

본 실험에 의하여 개발된 피부질환 치료제는 단순히 아토피뿐만 아니라 원인이 밝혀지지 않은 세포 내 기생충에 의한 피부질환에도 효능이 나타남.

The skin disease treatment developed through this experiment is effective not only for atopic dermatitis but also for skin diseases caused by intracellular parasites of unknown cause.

## 실시예 2 황(Sulfur) 성분의 살균제를 포함한 피부 질환 치료용 조성물 제조

## Example 2 Preparation of a composition for treating skin diseases containing a sulfur-containing disinfectant

### 유황이온액 제조

### Sulfur ion liquid manufacturing

2리터의 비이커에 수산화나트륨(NaOH) 200g과 황(S) 250g을 넣고 물($H_2O$) 500ml를 넣고 나무주걱이나 유리막대로 저어주었다. 발열반응에 의하여 120도에서 황이 녹기 시작하여 완전용해가 일어나면 물($H_2O$) 380ml를 추가로 넣어주고 식혔다.

Put 200g of sodium hydroxide (NaOH) and 250g of sulfur (S) in a 2-liter beaker, add 500ml of water ($H_2O$), and stir with a wooden spatula or glass rod. Sulfur began to melt at 120 degrees due to an exothermic reaction, and when complete dissolution occurred, an additional 380ml of water ($H_2O$) was added and cooled.

이 용액에 물을 추가하여 최종 1리터가 되게 하였다. (황 250g / 1,000ml)

Water was added to this solution to make a final volume of 1 liter. (Sulfur 250g / 1,000ml)

**유기규소이온액과 유황이온액의 반응**

**Reaction between organosilicon ionic liquid and sulfur ionic liquid**

유기규소이온액 500g의 pH를 L-arginine을 이용하여 10.0으로 조절하고, 유황이온액 2.5ml를 추가하여 10분간 교반하였다.

The pH of 500g of the organosilicon ion solution was adjusted to 10.0 using L-arginine, and 2.5ml of the sulfur ion solution was added and stirred for 10 minutes.

**총 구성**: 유기규소이온액 500g, 유황이온액 2.5ml, EGF 1ppm 용액 7.5g, FGF 1ppm 용액 7.5g, IGF 1ppm 용애 7.5g을 유효성분으로 포함하고, 부가 첨가물로 라루릴글루코사이드(유화제) 0.75g, 비타민 B3 2.5g, 글루타치온 0.75g을 더 첨가 희석하여 피부질환 치료용 조성물을 제조하였다.

**Total composition**: Contains 500g of organosilicon ion solution, 2.5ml of sulfur ion solution, 7.5g of EGF 1ppm solution, 7.5g of FGF 1ppm solution, and 7.5g of IGF 1ppm solution as active ingredients, and 0.75g of lauryl glucoside (emulsifier) as an additional additive., 2.5g of vitamin B3 and 0.75g of glutathione were further added and diluted to prepare a composition for treating skin diseases.

유황이온액은 유기규소이온액에 제일 먼저 투입하여 교반하고 여과하였다.

The sulfur ion solution was first added to the organic

silicon ion solution, stirred, and filtered.

이 조성물은 화장료로 사용할 수 있는 농도의 조성물에 해당한다.

This composition corresponds to a composition with a concentration that can be used as a cosmetic.

최종 pH는 L-arginine을 사용하여 10.0 이상으로 조절하였다.

The final pH was adjusted to over 10.0 using L-arginine.

pH가 낮을 경우 유황이온이 분해될 수 있기 때문이다.

This is because if the pH is low, the sulfur ions may break down.

## 적용례 6 중증 아토피 환자 – 황(Sulfur) 성분을 포함한 피부질환 치료용 조성물 적용례

## Application example 6 Patients with severe atopy – Example of application of a composition for treating skin diseases containing sulfur ingredients

**적용대상**: 30대 남성 중증 아토피 환자(심한 아토피로 듀피젠트를 2년간 주사하였어도 얼굴 입술 주변에 아토피가 치료되지 않을 정도로 매우 심한 아토피 환자)

**Applicable to**: Male patients in their 30s with severe atopic dermatitis. (A patient with severe atopic dermatitis, to the extent that the atopy around the face and lips was not cured even after two years of injections with Dupixent)

중증 아토피환자이기 때문에 세포 내 기생세균이 많아 치료기간이 6개월 이상 걸릴 것으로 예상해서 기간을 단축하기 위해서 결정

Because the patient is a patient with severe atopic dermatitis, there are many intracellular parasitic bacteria, so the treatment period is expected to take more than 6 months, so the decision was made to shorten the period

**적용방법**: 황 성분이 포함된 피부질환 치료용 조성물을 가려운 환부에 발라서 가려움증이 8시간 이후에 다시 지속되는지 관찰. 가려움증이 반복되지 않으면 추가적으로 도포하지 않고, 반복되

는 피부에만 도포하는 방식으로 사용

**How to apply**: Apply a composition for treating skin diseases containing sulfur to the itchy affected area and observe whether the itching continues again after 8 hours. If itching does not recur, do not apply additionally, but only apply to the skin where it recurs

**적용결과**:

**Application results**:

황 성분이 포함된 치료제를 바른 이후에는 가려움증이 해소되었으며 몇 시간 후 다시 가려움증이 시작되곤 하였다. 황 성분이 포함된 치표제를 바르는 중에 심한 통증이 발생하였다.

After applying a treatment containing sulfur, the itching was relieved, but the itching would begin again a few hours later. Severe pain occurred while applying dental treatment containing sulfur.

심한 통증은 엔도톡신 반응에 의해 숙주세포가 죽는 과정에서 오는 것으로 추정되었다.

It was assumed that severe pain came from the process of host cell death due to an endotoxin reaction.

균이 사멸하면, 숙주세포가 같이 사멸하고 그로 인해 염증이 발생하여 가려움증이 쉽게 가라앉지 않는 현상이 발생하였다.

When the bacteria die, the host cells also die, which causes inflammation and causes the itching to not subside easily.

염증반응이 심한 피부에 균과 숙주세포의 사멸로 인해 생긴 염증을 사혈(부항)으로 소량의 피를 제거해주면 가려움증이 해소되었다.

When a small amount of blood was removed through bloodletting (cupping) to remove inflammation caused by the death of bacteria and host cells on skin with a severe inflammatory response, the itching was relieved.

사혈(부항) 행위는 피부 가까운 곳에서 과량으로 사멸한 균으로 인해 발생한 염증성 물질 제거에 탁월하고 이로 인한 가려움증 해소에 도움이 된다.

Bloodletting (cupping) is excellent for removing inflammatory substances caused by excessively dead bacteria near the skin and helps relieve itching.

황 성분이 포함된 치료제는 아토피 피부염이 없는 깨끗한 피부에서는 반응하지 않았다.

Treatments containing sulfur did not react on clean skin without atopic dermatitis.

본 기술의 실시예 및 적용례를 통하여 본 바와 같이, 공통적으로는 피부질환 치료용 조성물을 피부에 발랐을 때, 사용자들이 즉시 또는 1분 이내에 아토피 피부염으로 인한 가려움증을 느끼지 않을 정도의 빠른 진정 효과를 보이며, 지속적으로서 사용 시 가려움증을 일으키는 부위가 점차 줄어들어 치료의 결과를 보이고 있다.

As seen through examples and application examples of the present technology, in common, when the composition for treating skin diseases is applied to the skin, it shows a rapid soothing effect to the extent that users do not feel itching due to atopic dermatitis immediately or within 1 minute. With continued use, the area causing itching gradually decreases, showing results of treatment.

본 기술에 의한 피부질환 조성물은 아토피증후군의 대표적인 증상인 아토피 피부염뿐만 아니라 아직 원인이 명확하지 않은 화폐상 습진에도 완화 내지 치료효과를 보이는 것으로 밝혀졌다.

It has been found that the skin disease composition according to the present technology has an alleviating or therapeutic effect not only on atopic dermatitis, which is a representative symptom of atopic syndrome, but also on nummular eczema, the cause of which is not yet clear.

아토피 피부염 사용사례에서 다음 두 가지는 매우 중요한 시사점을 제공한다.

In the atopic dermatitis use case, the following two provide very important implications.

그 첫째는 바르자마자 혹은 몇 분 안에 가려움증이 해소되는 것이고, 그 둘째는 딱지(scab)가 생긴다는 것이다. 이 두 가지 현상은 마이코플라즈마균이 치료제에 의해 사멸한다는 증거이다.

The first is that the itching is relieved as soon as applied or within a few minutes, and the second is that scabs form. These two phenomena are evidence that mycoplasma bacteria are killed by the treatment.

# 엔도톡신
## Endotoxin

엔도톡신(Endotoxin)이란 영문 뜻으로 알아보면 내(Endo) 독소(toxin)로 세균의 내부에 있는 독소라는 뜻이다.

Endotoxin, in English, is an endotoxin, which means a toxin inside bacteria.

세균이 일부러 방출하는 독소가 아닌 세균이 죽을 때 노출하게 되는 물질로 인체 내에서 독소로 작용한다.

It is not a toxin intentionally released by bacteria, but a substance exposed to bacteria when they die, and acts as a toxin in the human body.

이 현상으로 세균이 죽으면서 따가움, 통증을 유발하며, 숙주세포가 같이 죽어 딱지(scab)가 되는 것이다.

This phenomenon causes stinging and pain as the bacteria die, and the host cells also die and form scabs.

유황이온이 포함된 소독제의 통증의 강도는 세포 내 기생충의 양에 비례한다. 심한 경우 출산의 고통과 비견되기도 한다.

The intensity of pain from disinfectants containing sulfur ions is proportional to the amount of intracellular parasites. In severe cases, it can be compared to the pain of childbirth.

그러나, 유황이온이 없거나 소량 들어가 있는 세포 성장인자 위주의 치료제는 약간 따끔거리는 정도이다.

However, treatments that focus on cell growth factors that contain no or small amounts of sulfur ions only cause a slight tingling sensation.

지금까지 아토피 피부염 치료제들은 아토피 피부염 부위에 딱지(scab)를 발생시키지 않았다.

Until now, atopic dermatitis treatments have not caused scabs to form on the atopic dermatitis area.

그러나 금번 기술의 피부질환 치료용 조성물은 반드시 딱지(scab)를 발생시키고 있으며, 이는 엔도톡신 현상에 의한 것이라 볼 수 있다.

However, the composition for treating skin diseases of this technology always generates scabs, which can be seen as being caused by the endotoxin phenomenon.

즉, 딱지(scab)는 엔도톡신 현상으로 세포 내 세균이 숙주세포와 함께 사멸했다는 증거라 할 수 있다. (이로 인하여 태선화 현상이 발생하지 않는다.)

In other words, scabs can be said to be evidence that intracellular bacteria have died along with the host cells due to the endotoxin phenomenon. (Because of this, lichenification does not occur.)

본 기술의 조성물은 막수송체 역할을 규소이온이 한다는 것을 밝혀냈고, 황(Sulfur) 성분의 살균제 및 세포 성장인자를 사용하여 아토피 치료에 진전을 보였다는 점에서 의의가 있다.

The composition of this technology is significant in that it was discovered that silicon ions play the role of a membrane transporter, and progress was made in the treatment of atopy by using sulfur-containing disinfectants and cell growth factors.

# 부록 1 매일경제(www.mk.co.kr) 신문 기사내용 및 영문번역

# *Appendix 1* Maeil Business Newspaper (www.mk.co.kr) newspaper article content and English translation

**㈜에이치앤비나인 바이오 연구소, '마이코플라즈마' 세포 간 증식 기작 규명**

**HnB9 Bio Research Institute, Identifies the Intercellular Proliferation Mechanism of 'Mycoplasma'**

[입력(input): 2019-10-04 13:50:59]

국내 바이오 기업인 ㈜에이치앤비앤나인(대표 유재덕)이 피부섬유아세포와 암세포에서 발생하는 '마이코플라즈마(Mycoplasma)'의 면역세포 회피를 위한 세포 간 이동 · 증식 기작을 규명한 내용을 국제 학술지인 'BMB Reports'에 게재해 학계 주목을 받고 있다.

HnB9 Co., Ltd. (CEO Jae-deok Yoo), a domestic bio company, published its findings on the cell-to-cell migration and proliferation mechanism of 'Mycoplasma', which occurs in skin fibroblasts and cancer cells, to evade immune cells, in the international academic journal 'BMB'. It is receiving attention from academic circles after being published in 'Reports'.

마이코플라즈마는 세균(Bacteria)의 일종으로 일부는 병원성이 좋아 폐렴(Atypical pneumonia), 피부건선(psoriasis), 아토피(Atopic dermatitis), 류마티스 관절염(rheumatoid arthritis), 다발성경화증(multiple sclerosis), 피부암(melanoma)과 같은 질환을 유발시킨다.

Mycoplasma is a type of bacteria, and some of them are highly pathogenic and cause pneumonia, psoriasis, atopic dermatitis, rheumatoid arthritis, multiple sclerosis, and skin cancer causes diseases such as.

특히 바이러스와 세균의 중간 성질을 가진 미생물로 일반 세균과 다르게 세포벽이 없어 페니실린, 벵타-락탐과 같은 항생제에 내성이 강한 것으로 알려져 있다.

In particular, it is a microorganism with properties intermediate between viruses and bacteria, and unlike ordinary bacteria, it has no cell wall and is known to be highly resistant to antibiotics such as penicillin and bengta-lactam.

㈜에이치앤비나인은 마이코플라즈마 감염과 피부질환 발병의 연관성을 연구해 오던 중 마이코플라즈마가 면역세포의 공격을 피해 섬유아세포의 안으로 침입하여 생존할 수 있으며, 세포와 세포 사이에 가늘고 긴 터널(tunneling nanotube)을 생성시켜 이동한 후 증식한다는 것을 국내외 최초로 규명해냈다.

HnB9 Co., Ltd. has been researching the relationship between mycoplasma infection and the development of skin diseases, and has discovered that mycoplasma can survive by invading fibroblasts to avoid attack by immune cells, and that a long, thin tunnel (tunneling nanotube) exists between cells. ) was created, moved, and then proliferated for the first time domestically and internationally.

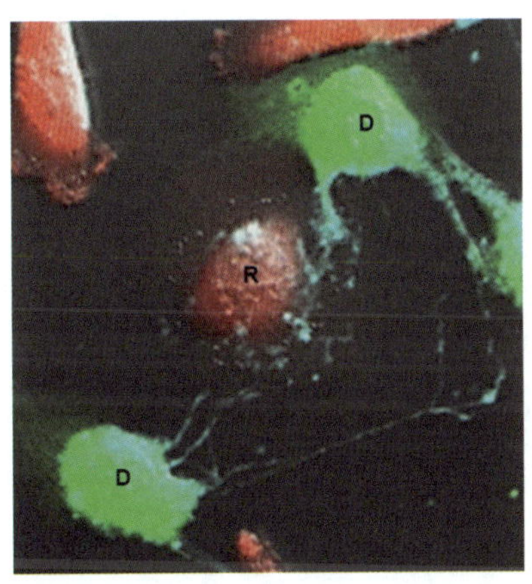


**감염된 세포(D, 녹색)로부터 미세터널을 통해 미감염 세포(R, 붉은색)로 미코플라즈마(하늘색)가 이동하는 모습**
**Mycoplasma (light blue) moving from infected cells (D, green) to uninfected cells (R, red) through a microtunnel**

현재 ㈜에이치비앤나인은 위와 같은 연구성과를 바탕으로 피부 건선과 아토피, 피부암의 예방과 완화를 위한 기능성 화장품을 개발하고 있으며, 피부질환 외에도 폐렴, 다발성 경화증, 류마티스 관절염, 피부암(흑색종) 등 마이코플라즈마 유발 질환을 개선할 수 있는 의약품 원료 개발도 함께 집중할 예정이다.

Currently, HnB9 is developing functional cosmetics

to prevent and alleviate skin psoriasis, atopy, and skin cancer based on the above research results. In addition to skin diseases, it is also used to treat mycoplasma such as pneumonia, multiple sclerosis, rheumatoid arthritis, and skin cancer (melanoma). We also plan to focus on developing pharmaceutical raw materials that can improve causing diseases.

유재덕 대표는 "대부분의 바이러스나 세균은 우리 몸 속에 침입하게 되면 생체 내 면역세포에 의해 죽거나 사라지지만 마이코플라즈마는 그렇지 않다"면서 "이번 연구를 통해 마이코플라즈마가 우리 몸 속 면역시스템을 회피하기 위해 세포 안으로 침입 후 세포와 세포 사이에 미세터널을 만들어 증식하게 된다는 것을 밝혀냈다"고 밝혔다.

CEO Jae-deok Yoo said, "When most viruses or bacteria invade our bodies, they are killed or disappeared by immune cells in the body, but this is not the case with mycoplasma." "We found that after invading, cells proliferate by creating microtunnels between cells."

한편, ㈜에이치앤비나인은 토탈 바이오 기업으로 도약을 위해 김재환 박사(煎 서울대학교 의학연구원 연구교수, 煎 MD Anderson Cancer Center 연구원), 김봉우 박사(고려대학교 연구교수), 임지헌 박사(煎 삼성의료원 줄기세포 재생의학 연구소 책임연구원) 등 소장급 연구원 3명을 올해 초 영입하며 기업부설연구소를 확장 및 세분화하는 등 바이오 독자 기술을 기반으로 화장품 및 의약품 원료 개발 분야에서 크게 두각을 나타내고 있다.

Meanwhile, in order to leap forward as a total bio company, HnB9 Co., Ltd. has teamed up with Dr. Jaehwan Kim (research professor at Seoul National University Medical Research Institute, researcher at MD Anderson Cancer Center), Dr. Bongwoo Kim (research professor at Korea University), and Dr. Jiheon Lim (currently Samsung Medical Center). By recruiting three director-level researchers, including the chief researcher of the Stem Cell and Regenerative Medicine Research Institute, early this year, and expanding and subdividing the company-affiliated research institute, the company is making a mark in the field of developing cosmetics and pharmaceutical raw materials based on its proprietary bio technology.

# 부록 2 비염환자의 아토피 발현 및 치료 사례

# Appendix 2 Cases of atopy presentation and treatment in rhinitis patients

"이 치료사례는 아토피증후군 중 비염환자에게서 아토피 발현 사례를 통해 아토피증후군의 질병이 같은 원인에 의하여 발생한다는 것을 증명하는 효과가 있는 중요한 사례입니다."

"This treatment case is an important case that has the effect of proving that the disease of atopic syndrome is caused by the same cause through a case of atopy in a patient with rhinitis."

특히 마이코플라즈마에 의하여 아토피가 발현되는지에 대한 강력한 증거로 작용될 사진자료가 첨부됩니다.

In particular, photographic data is attached that will serve as strong evidence as to whether atopy is caused by mycoplasma.

또한 치료과정을 면밀하게 관찰하고, 환자의 의견을 청취함으로서 치료방법을 연구하는 데 도움을 주기 위한 사례입니다.

In particular, this is an example to help research treatment methods by closely observing the treatment process and listening to the patient's opinions.

## 환자의 치료 전 기초자료
## Patient' s basic data before treatment

1. 치료 전 환자는 외견상 지간신경종(Morton's Neuroma)의 질병을 앓고 있었다. 이런 이유로 다리를 절뚝거리는 상태였다.
1. Before treatment, the patient appeared to be suffering from Morton's Neuroma. For this reason, he was limping.

2. 호흡기는 비염(축농증)이 심한 상태였으며, 인터뷰에 의해 알아낸 것은 8살 때부터 비염이 있었기 때문에 약 52년간 비염을 앓고 있었다.
2. The respiratory system was suffering from severe rhinitis (sinusitis), and what we found out through the interview was that he had had rhinitis since he was 8 years old, so he had had it for about 52 years.

3. 눈에는 약간의 결막염 증상을 보인 상태였다.

3. The eyes showed some symptoms of conjunctivitis.

4. 육안상으로 아토피 피부염은 없었다.
4. Visually, there was no atopic dermatitis.

5. 아토피 발현 후 피부 상태(2023년 10월 25일 아토피 피부염 발현)
5. Skin condition after atopic dermatitis occurs (Appearance of atopic dermatitis on October 25, 2023)

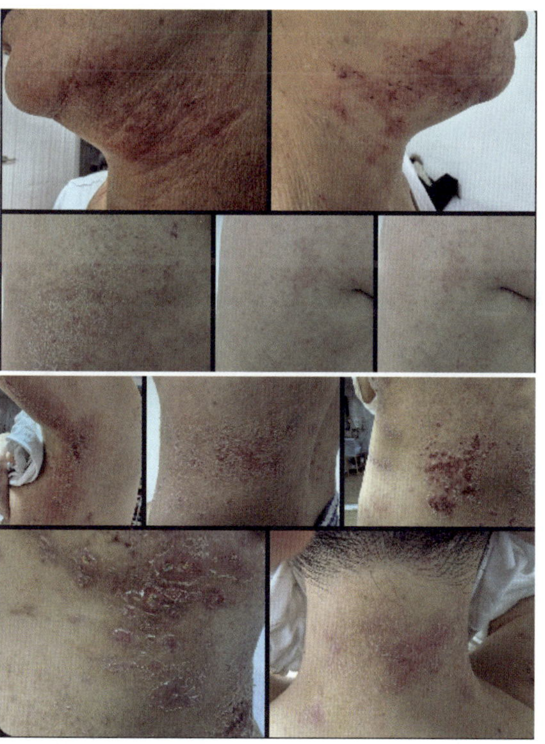

몸 속에 숨어있는 아토피 원인균의 피부염 발현은 두 가지 방법으로 이루어졌다.

Dermatitis caused by atopy-causing bacteria hidden in the body can occur in two ways.

첫째는 꾸준한 규소이온의 섭취다. (1일 30ml 섭취, 6개월~1년) 규소이온의 섭취는 세포 속에 숨어있는 아토피 원인균을 세포 바깥으로 몰아내는 역할을 한다.

The first is the steady intake of silicon ions. (Intake of 30ml per day, 6 months to 1 year) Ingestion of silicon ions plays a role in expelling atopic dermatitis-causing bacteria hidden in cells out of the cells.

원인은 명확하게 밝혀지지 않았지만, 규소이온이 세포막수송체 이외에 세포에 영양분으로 작용하는 것으로 추정된다.

Although the cause is not clearly known, it is presumed that silicon ions act as nutrients for cells in addition to being a cell membrane transporter.

즉 세포에 영양을 공급하여 세포 내 기생충을 몰아내는 현상으로 보인다.

In other words, it appears to be a phenomenon that supplies nutrients to cells and drives out intracellular parasites.

두 번째는 상재균처럼 숨어있는 세포 내 기생충을 막수송체인 규소이온에 의하여 수송된 유황이온과 세포 성장인자로 인하여 치료되는 과정에서 세균의 저항으로 인하여 아토피 피부염 증세가 발현된다.

Second, in the process of treating hidden intracellular parasites like common bacteria with sulfur ions and cell growth factors transported by silicon ions, which are membrane transporters, symptoms of atopic dermatitis appear due to bacterial resistance.

주변의 수없이 많은 아토피 증후군 환자들을 상대로 아토피의 발현 여부를 증명할 수 있다. (비교적 안전한 실험에 해당한다. 치료방법이 생겼으므로…)

The presence of atopic dermatitis can be proven against countless atopic syndrome patients around us. (It is a relatively safe experiment. Now that a treatment method has been found…)

## 첫 번째 조치사항과 그 결과
First actions and results

1. 지간신경종의 회복을 위하여 섭취 가능한 유기규소이온액(한국 상품명 K8)을 6개월간 섭취하도록 하면서 매주 조금씩 운동량을 늘려나갔다.
1. To recover from Morton's Neuroma, I took organosilicon ions (Korean brand name K8), an ingestible mineral, for 6 months and gradually increased the amount of exercise each week.

2. 결과: 통증이 완화가 되어 절뚝거리지 않고 정상적인 발걸음을 하게 되었고, 하루 5,000보 정도의 운동이 가능하였다.
2. Result: The pain was relieved, the patient was able to walk normally without limping, and was able to exercise about 5,000 steps a day.

3. 예상치 못한 현상: 하반신에 가려움증이 발생하기 시작하여 규소이온 복용한 지 6개월 되었을 때부터 참기 어려울 정도로 가려움증이 심해졌다.
3. Unexpected phenomenon: Itching began to occur in the lower body, and after 6 months of taking silicon ion, the itching became unbearable.

## 예상치 못한 현상에 대한 조치사항과 그 결과
Measures and results for unexpected phenomena

1. 지간신경종(Morton's Neuroma)이 완화되어 규소이온 K8의 섭취를 중단하고, 가려움증의 원인을 살펴보았으나 나아지지 아니하였다.
1. Morton's Neuroma was alleviated and the intake of silicon ion K8 was stopped and the cause of the itching was investigated, but there was no improvement.

2. 유기규소이온액(킬레이트 실리콘이온) 한국 상품명 K9 500g에 EGF, FGF, IGF 1ppm 용액 10ml씩 총 30ml를 첨가하여 피부에 바르는 화장품을 제조하여 6개월간 사용케 하였다. (약간의 계면활성제를 추가하여 퍼짐성을 높였다.)
2. Organic silicon ions (chelated silicon ions) A total of 30ml of 10ml of EGF, FGF, and IGF 1 ppm solution were added to 500g of K9, a Korean brand name, to prepare cosmetics to be applied to the skin and used for 6 months. (Spreadability was improved by adding a little surfactant.)

3. **결과: 하반신의 가려움증이 급격하게 개선되어 최종적으로 발생 6개월 만에 가려움증이 완치되었다. (작은 딱지들이 생기면서 가려운 부위가 줄어들다가 가려움증이 없어졌다.)**

3. **As a result: the itching of the lower body improved rapidly, and the itching was finally cured 6 months after the onset. (As small scabs formed, the itchy area decreased and then the itching disappeared.)**

4. **예상치 못한 현상: 유기규소이온액에 희석하지 않은 EGF, FGF, IGF 등 성장인자는 피부에 사용했을 때는 가려움증을 줄여주는 효과가 거의 없었다. 그러나, 규소이온액에 희석한 경우에는 피부에 바르는 즉시 혹은 수분 이내에 가려움증을 잡아주는 효과가 나타났다.**

4. **Growth factors such as EGF, FGF, and IGF that were not diluted in organosilicon ion solution had little effect in reducing itching when used on the skin. However, when diluted with silicon ion solution, it was effective in suppressing itching immediately or within a few minutes of application to the skin.**

이 현상은 기존에 알려져 있는 '규소가 칼슘을 뼈로 운반하는 화물차와 같은 역할'을 하듯이 규소이온이 막수송체 역할을 하여 성장인자를 세포 속으로 운반하여 나타난 현상으로 해석되었다.

This phenomenon was interpreted as a phenomenon in

which silicon ions act as membrane transporters to transport growth factors into cells, just as silicon acts like a truck transporting calcium to bones.

## 세 번째 조치사항과 그 결과
Third action and its results

1. 지간신경종(Morton's Neuroma)의 완치를 위하여 다시 유기규소이온액(한국 상품명 K8)을 1일 30ml씩 섭취하고 걷기 운동량을 늘려갔다.
1. To completely cure Morton's Neuroma, I again consumed 30ml of organosilicon ionic liquid (Korean brand name K8) per day and increased the amount of walking exercise.

2. 해당 유기규소이온액을 섭취하면서 가려움증이 다시 올라오기 시작하면서 전신이 중증아토피 현상으로 번지기 시작하였다.
2. After ingesting the organosilicon ion solution, the itching began to return and the entire body began to develop severe atopic dermatitis.

3. 아토피증후군에는 아토피 피부염, 알레르기성 결막염, 천식, 알레르기성 비염이 포함되는데 8살 때부터 비염을 가지고 있던 환자의 몸 속에 마이코플라즈마균이 증식한 상태로 상재균처럼 있다가 규소이온액을 섭취함으로 인하여 규소이온액이 막수송체로서 이온채널을 여는 역할을 하는 과정에서 몸 속에 있던 마이코플라즈마 균들이 습성에 맞게 피부 쪽으로 이동하

기 시작한 것으로 추정하였다.

3. Atopic syndrome includes atopic dermatitis, allergic conjunctivitis, asthma, and allergic rhinitis. Mycoplasma bacteria proliferate in the body of a patient who has had rhinitis since the age of 8, and grows as a common bacteria and then ingests silicon ion liquid. Therefore, it was assumed that the mycoplasma bacteria in the body began to move toward the skin according to their habits as the silicon ionic liquid played the role of opening the ion channel as a membrane transporter.

4. 지속적인 유기규소이온액의 섭취는 몸 속에 숨어있던 마이코플라즈마균들을 피부 쪽으로 이동시키는 현상을 유발하였다.

4. Continuous intake of organosilicon ionic liquid caused the mycoplasma bacteria hidden in the body to move toward the skin.

## 5. 치료에 사용된 물질(Substances used in treatment)

### 1) 소독용 화장품(Disinfection cosmetics)

(1) 규소이온 기반 종합미네랄(한국 상품명 미네랄 밸런스) 30ml

(1) Silicon ion-based comprehensive mineral (Korean

brand name: Mineral Balance) 30ml

(2) 화장품 원료 유황이온액 100ml

(2) Cosmetic raw material sulfur ion solution 100ml

(3) 구입처: www.si-ion.com

(3) Where to purchase: www.si-ion.com

(4) 유황이온액 30ml + 미네랄 밸런스 30ml를 희석하여 소독용 화장품을 제조하였다.

(4) Disinfecting cosmetics were prepared by diluting 30ml of sulfur ion solution + 30ml of mineral balance.

**2) 세포 성장인자를 포함하는 기초화장품(이하 '성장인자 화장품'이라 한다.)**

**2) Basic cosmetics containing cell growth factors (hereinafter referred to as 'growth factor cosmetics')**

| 투입순서 | 원재료명<br>Raw material name | 소요량<br>Amount |
|---|---|---|
| Input order | 규소이온 K9<br>Chelated silicon ionic liquid K9 | 500.00g |
| 1 | EGF 1ppm(10ml) | 7.50g |
| 2 | FGF 1ppm(10ml) | 7.50g |

| 투입순서 | 원재료명<br>Raw material name | 소요량<br>Amount |
|---|---|---|
| 3 | IGF 1ppm(10ml) | 7.50g |
| 4 | MSM | 7.50g |
| 5 | 라우릴글루코사이드(유화제)<br>Lauryl Glucoside (emulsifier) | (약간) 0.25g<br>one drop |
| 6 | 아르기닌(pH 조절제)<br>L-arginine(pH adjuster) | 20.0g |
| 7 | 제조총량<br>Total manufacturing quantity | 585.6g |

### 3) 밤에 사용하는 방법

### 3) How to use at night

'성장인자 화장품' 70ml에 소독용 화장품 100~200방울을 첨가하여 피부에 사용(L-아르기닌을 추가하면 유황이온이 분해되는 것을 방지할 수 있다.)

Add 100 to 200 drops of disinfectant cosmetics to 70ml of 'growth factor cosmetics' and use on the skin(adding L-arginine can prevent sulfur ions from decomposing.)

### 4) 소독용 화장품의 사용법(주의: 심한 통증)

### 4) How to use disinfectant cosmetics (caution: severe pain)

일주일에 1회 10분 이상 1시간 이내 얼굴을 제외한 신체 전체에 골고루 펴 바름. (이때 피부에 균이 있는 부분에만 작용하여 살균현상이 일어나며 통증을 유발함. 정상 피부는 아무런 반응이 없음.)

Apply evenly over the entire body, excluding the face, once a week for at least 10 minutes and within 1 hour. (At this time, it only acts on the part of the skin where there are bacteria, causing sterilization and causing pain. Normal skin has no reaction.)

긴 시간 동안 소독을 할 경우 중간중간 덧바름.

If disinfecting for a long time, reapply occasionally.

원인균이 많을 경우 극심한 통증이 유발됨.

If there are many causative bacteria, severe pain may occur.

## 6. 기타사항

## 6. Other matters

피부병변 상태 사진을 설명하면서 추가사항을 기술함.

Describe additional information while explaining photos of skin lesions.

### 피부병변 사진 및 설명

### Photos and explanations of skin lesions

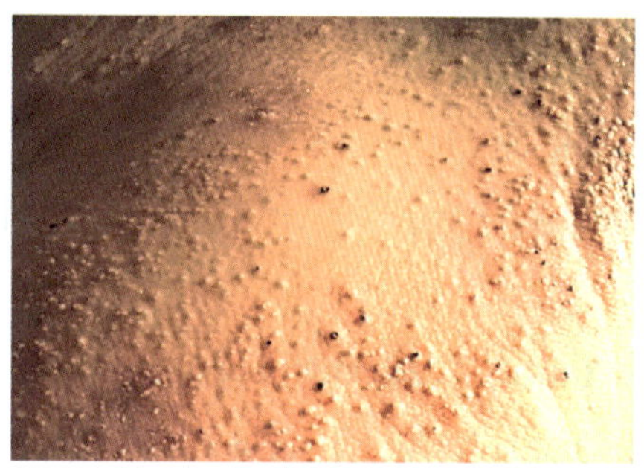

비염 및 지간신경종을 이유로 규소이온 섭취 후 발생한 전신 아토피의 초기 사진. 검은색 딱지와 하얀색 딱지가 혼재하여 있다.

The initial picture of whole-body atopy that occurred after ingestion of silicon ions due to rhinitis and interdigital neuroma shows black and white scabs mixed together.

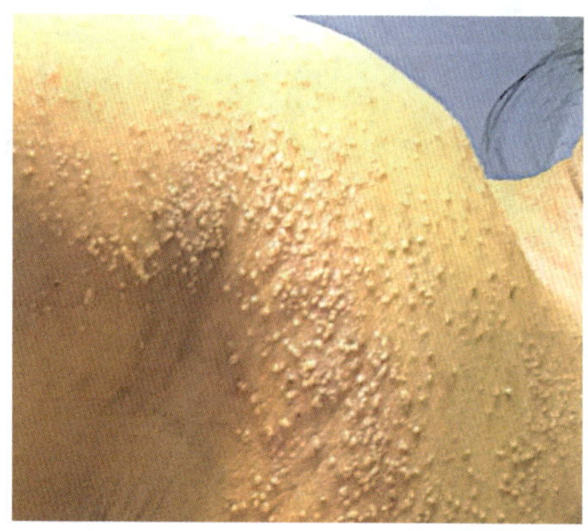

소독용 화장품으로 소독 중인 초기 사진(아토피 원인균들이 물집을 만들어 소독에 저항하고 있다.)

Initial photo of disinfection with disinfectant cosmetics (bacteria that cause atopy are forming blisters and resisting disinfection.)

**환자 의견**: 소독이 끝난 후 피부 바깥의 물집과 피부 속의 물집이 따로 있으며 해당 물집을 터뜨려주면 시원해지면서 가려움증이 해소됨.

**Patient's opinion**: After disinfection, there are separate blisters on the outside of the skin and blisters inside the skin. If you pop the blisters, you will feel cool and the itchiness will be relieved.

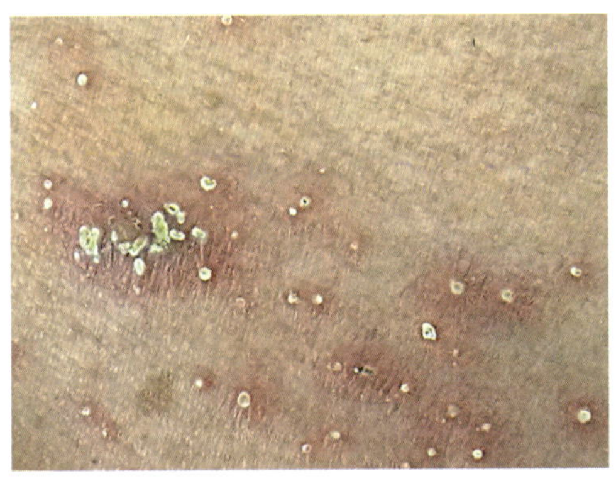

**아토피의 원인균이 소독으로 인하여 피부 위에 나타난 모습(소독 1시간 )**

**The bacteria causing atopic dermatitis appear on the skin due to disinfection. (1 hour of disinfection)**

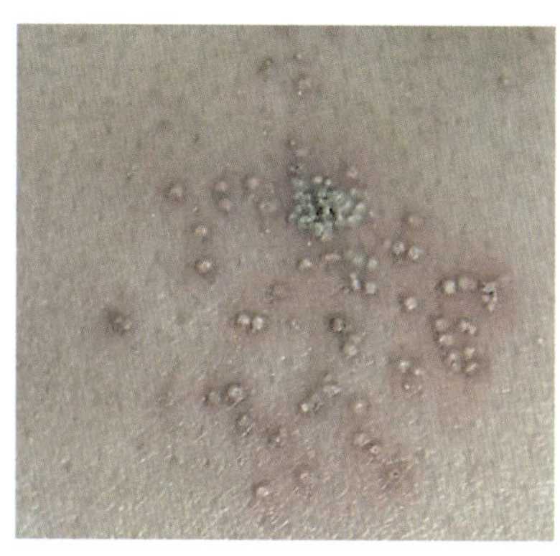

또 다른 형태의 아토피 원인균의 모습(소독 1시간)

Another type of atopy-causing bacteria (1 hour of disinfection)

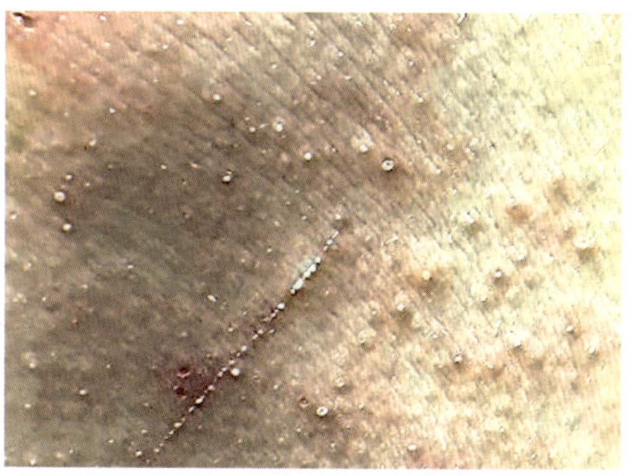

아토피 원인균이 옮겨 다니다가 살균 당한 흔적(매일경제 뉴스에서 본 마이코플라즈마 증식사진과 유사하다.) (소독 1시간)

Traces of atopy-causing bacteria moving around and being sterilized (similar to the photo of mycoplasma growth seen in Maeil Business News). (1 hour of disinfection)

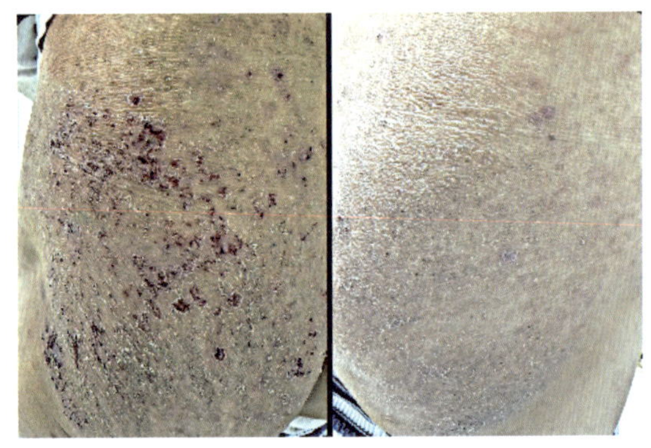

**둔부에 1시간 30분 이상 소독 후 피부 속에서 균들이 죽은 모습 (피부색이 검어졌지만 이 이후로 2개월째 더 이상 가려움이 발생하지 않으면서 피부가 정상으로 돌아왔다.)**

**After disinfecting the buttocks for more than 1 hour and 30 minutes, the bacteria in the skin were dead (the skin color became dark, but after 2 months, it no longer occurred, and the skin returned to normal.)**

**환자 의견**: 소독용 화장품으로 오랜 시간 소독할 경우 한꺼번에 많은 균이 죽기 때문에 피부에서 가깝게 사혈(부항)을 통해서 균이 죽으면서 발생시킨 염증물질을 빠르게 제거해줘야 가려움증이 즉시 해소됩니다. (염증성 물질이 많이 발생합니다.)

**Patient's Opinion**: When disinfecting with disinfectant cosmetics for a long time, many bacteria are killed at once, so itching is immediately relieved when the inflammatory substances generated by the bacteria are quickly removed through bloodletting (cupping) close to the skin. (A lot of inflammatory substances are generated.)

많은 양의 균을 소독해야 할 경우 피부에 바르는 방식 이외에 사용할 수 있는 항생제가 있었으면 좋겠습니다.

If you need to disinfect a large amount of bacteria, I wish there was an antibiotic that could be used other than applying it to the skin.

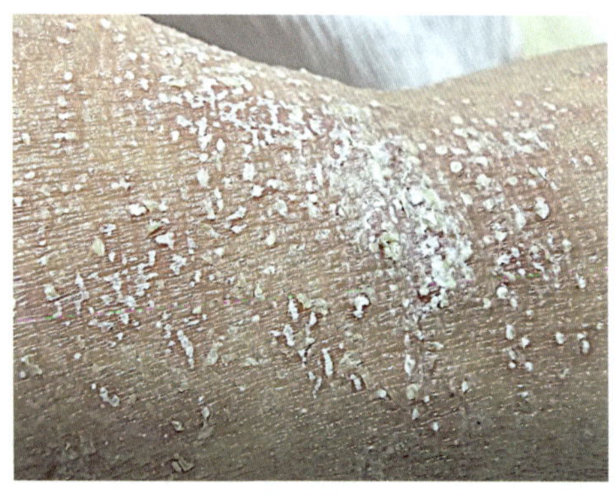

**긴 시간 소독 후 피부를 뚫고 나타난 아토피 원인균의 모습(소독용 화장품의 유황이온을 피해 급하게 피부 위쪽으로 균들이 뚫고 나온 모습을 보여주고 있다.)**

**Atopic dermatitis-causing bacteria that appeared through the skin after a long period of disinfection (This shows that the bacteria quickly broke through the upper part of the skin to avoid the sulfur ions of the disinfecting cosmetics.)**

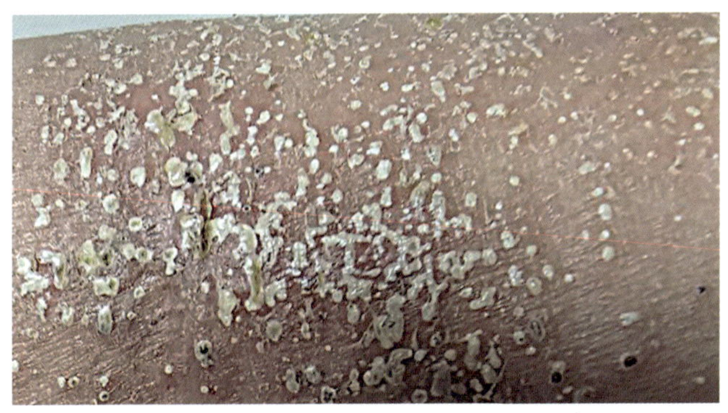

**아토피 원인균이 소독 후 말라가는 모습(흰색과 검은색의 딱지로 변하며 딱지가 모래알처럼 많이 생성됩니다.)**

**Atopy-causing bacteria dry up after disinfection (they turn into white and black scabs, and many scabs are formed like grains of sand.)**

**환자 의견**: 소독 후 딱지가 생기고 난 다음 딱지가 떨어질 때 안티프라민(한국에서 판매하는 피부연고)으로 마사지하듯 오랜 시간 발라주면 수분을 보충하고, 모래알처럼 많은 양의 딱지를 쉽게 제거할 수 있습니다.

**Patient's opinion**: After disinfection, when scabs form and the scabs fall off, apply antipramine (skin ointment sold in Korea) for a long time as if massaging, to replenish moisture and easily remove a large amount of scabs like grains of sand.

저는 소독 후 적당한 시간에 안티프라민을 사용하기 위하여 500g짜리 큰 용기의 안티프라민을 지속적으로 사용하였습니다. 안티프라민은 진통제 역할도 하고, 보습제 역할도 했습니다.

I continuously used a large 500g container of antipramine to use it at an appropriate time after disinfection. Antipramine acts as both a painkiller and a moisturizer.

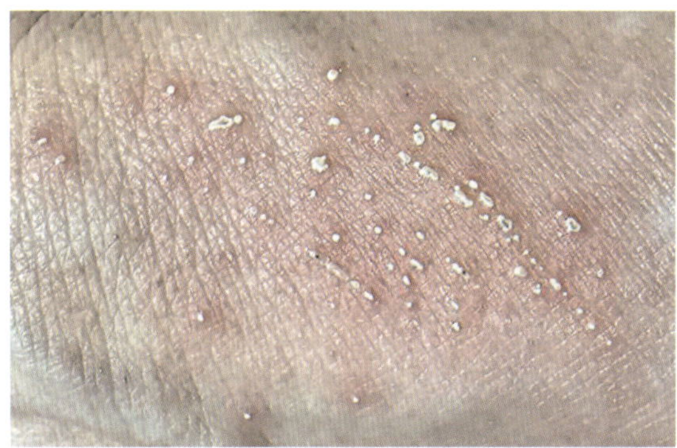

**피부 속에서 균이 이동 중에 살균되어 나타난 원인균의 사진(매일경제 뉴스에서 본 마이코플라즈마 증식사진과 유사하다.)**

**A photo of the causative bacteria appearing in the skin after being sterilized during movement (similar to the photo of mycoplasma growth seen in Maeil Business News(www.mk.co.kr).)**

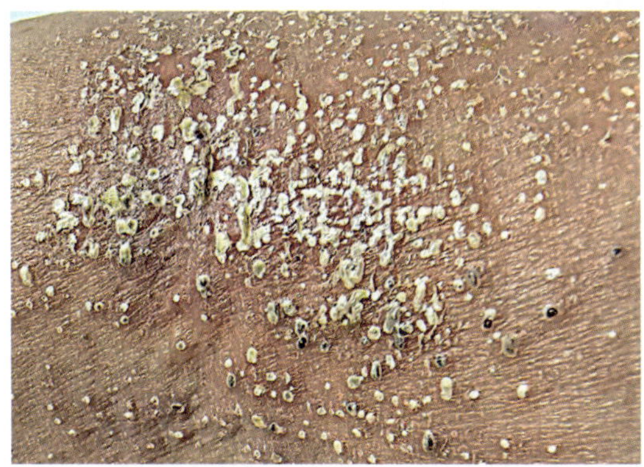

**원인균이 소독으로 인하여 피부 바깥으로 나온 모습**

**Appearance of causative bacteria coming out of the skin due to disinfection**

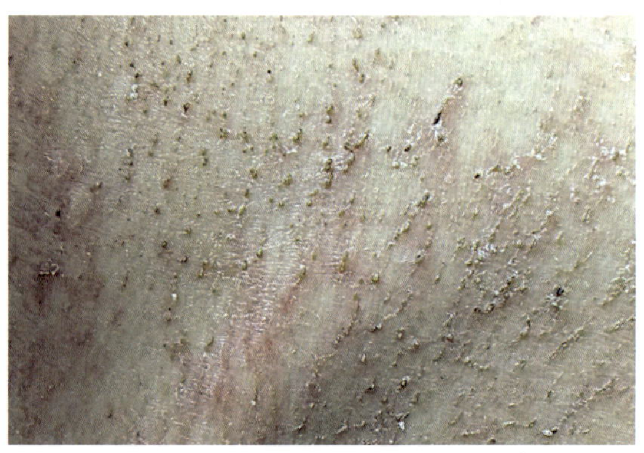

**균이 죽으면서 딱지가 형성된 모습(엔도톡신 현상에 의해 숙주세포가 같이 사멸하여 딱지를 형성한다. 이 딱지마저 독성이 있어서 약간의 가려움증과 따가움을 발생시킨다.)**

**A scab is formed as the bacteria die (host cells die together due to the endotoxin phenomenon and form a scab. This scab is toxic and causes slight itching and stinging.)**

**환자 의견**: 안티프라민으로 전신마사지를 하면 수분을 보충해주면서 모래알처럼 많은 딱지가 쏟아져 나옵니다.

**Patient's opinion**: When you massage your body with antipramine, it replenishes moisture and removes as many scabs as grains of sand.

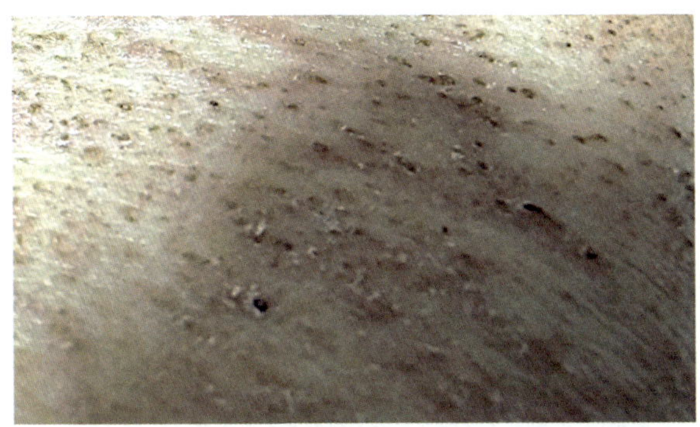

떨어지기 직전의 딱지의 모습(상처를 남기지 않고 깨끗하게 치유된다.)

Appearance of a scab just before it falls off (It heals cleanly without leaving a scar.)

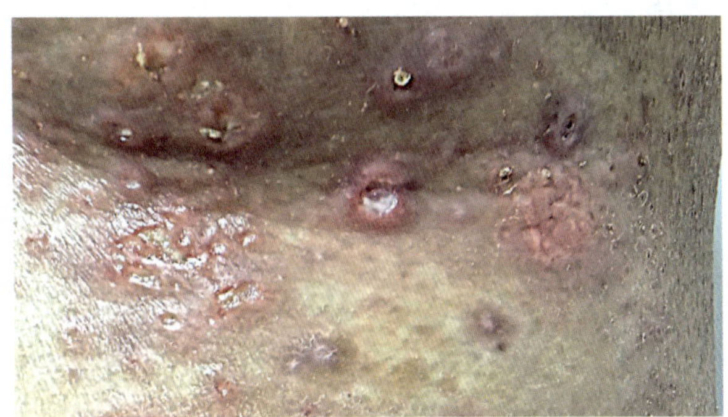

살이 접히는 부위에 생긴 심한 아토피 피부염 증상(살이 접히는 부위는 상대적으로 균이 많기 때문에 세심한 관리가 필요하다.)

Symptoms of severe atopic dermatitis in the area where the skin folds (the area where the skin folds has relatively many bacteria, so careful management is required.)

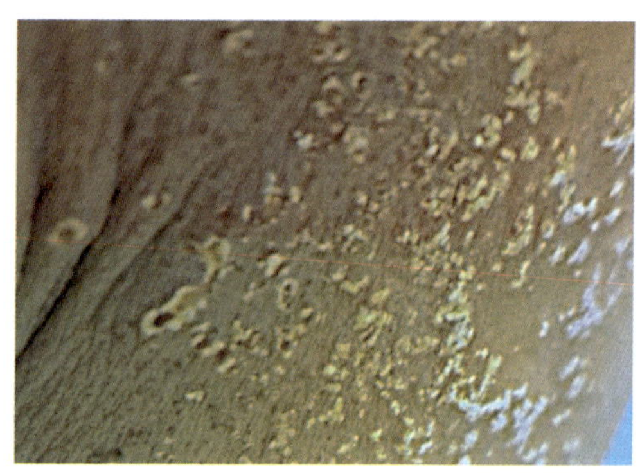

소독 중에 튀어나온 아토피 원인균의 모습(소독용 화장품을 만들어서 1시간 이상 심하게 소독을 할 경우 원인균이 소독성분을 피해 피부 바깥으로 튀어나온다. '심한 고통 주의')

Appearance of atopy-causing bacteria sticking out during disinfection (If you make disinfecting cosmetics and disinfect them thoroughly for more than an hour, the causative bacteria avoid the disinfecting ingredients and jump out of the skin. 'Beware of severe pain')

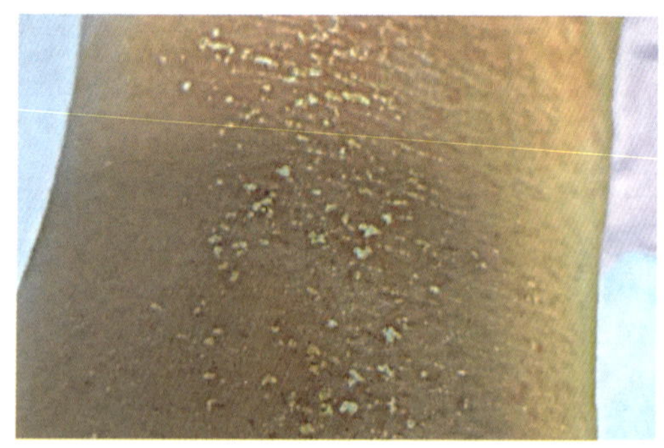

소독 중 튀어나온 아토피 원인균의 모습(마이코플라즈마균이 협

**기성 세균이므로 과산화수소 소독을 병행하면 도움이 된다.)**

**Appearance of atopy-causing bacteria sticking out during disinfection (Because mycoplasma bacteria are anaerobic bacteria, it is helpful to combine sulfur ion disinfection and hydrogen peroxide disinfection.)**

어깨 부위에 아토피 원인균이 이동하다가 소독되어 그 형태가 나타났다. (마이코플라즈마균의 이동방식과 동일하게 관찰된다.)

The bacteria causing atopic dermatitis moved to the shoulder area and were disinfected, resulting in its appearance. (It is observed in the same way as the movement method of mycoplasma bacteria.)

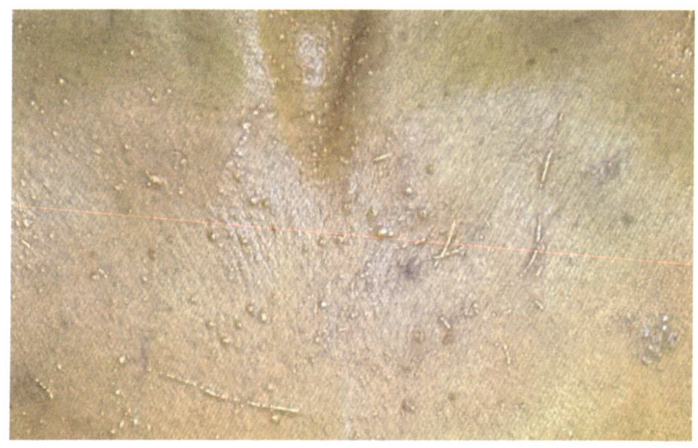

앞가슴 부위로 아토피 원인균들이 이동하다가 소독되어 그 형태가 나타났다. (마이코플라즈마균 이동사진과 동일하다.)

Atopic dermatitis-causing bacteria moved to the front chest area, were disinfected, and appeared in that form. (Same as the picture of mycoplasma bacteria movement.)

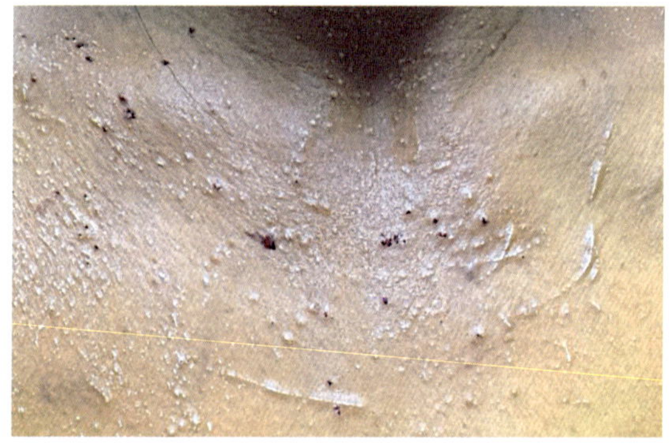

소독시간이 길어지면 균이 죽으면서 세포도 죽어 점성출혈이 일어난다. (소독시간 1시간 30분)

If the disinfection time is prolonged, the bacteria die and the cells die, causing viscous bleeding. (Disinfection time 1

hour and 30 minutes)

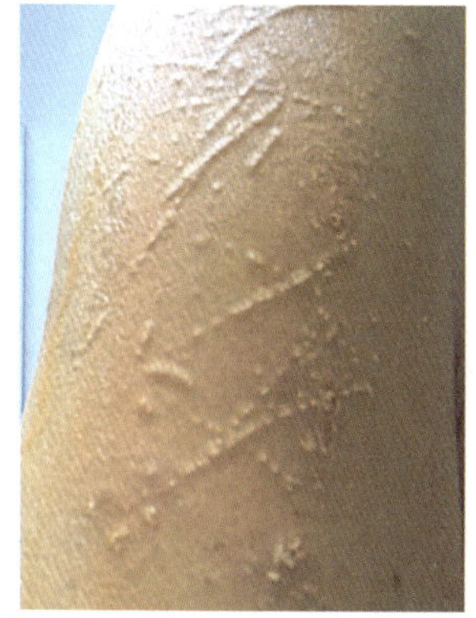

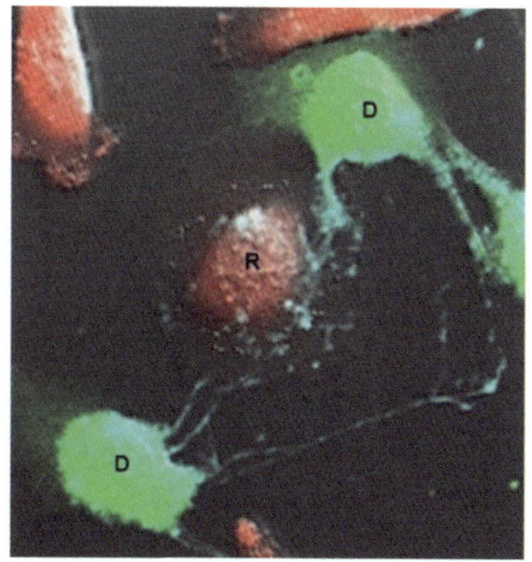


**마이코플라즈마균의 이동방식과 같은 유형으로 이동하다 소독되어 나타난 형태(우측은 마이코플라즈마균)**

**It appears after moving in the same way as mycoplasma bacteria (Mycoplasma bacteria on the right)**

소독용 화장품은 아토피 원인균을 육안으로 관찰할 수 있는 방안을 제시해주고 있다. (대신 극심한 고통을 참아야 한다.)

Disinfecting cosmetics provide a way to observe atopy-causing bacteria with the naked eye. (Instead, you have to endure extreme pain.)

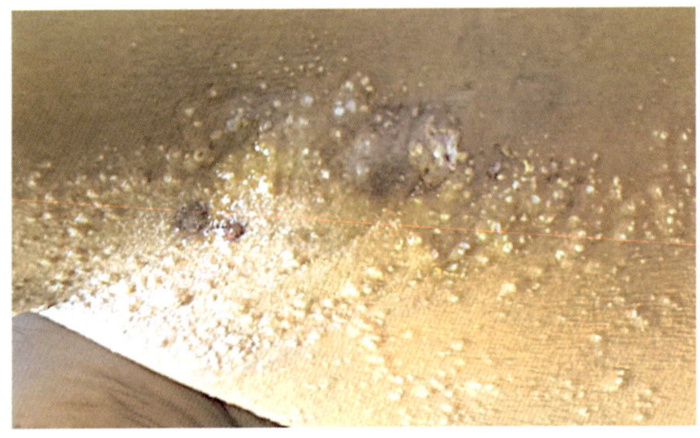

**균이 저항하는 모습(균은 피부 바깥쪽과 안쪽에 물집을 만들어서 소독에 저항한다.)**

**Bacteria resistance (Bacteria resist disinfection by creating blisters on the outside and inside of the skin.)**

**환자 의견**: 피부에 생긴 물집과 피부를 만졌을 때 피부 속에 생긴 물집을 제거해주어야 균이 죽으면서 가려움증이 해소됩니다.

**Patient's opinion**: Blisters that form on the skin and when the skin is touched must be removed to kill the bacteria and relieve itching.

세균이 사멸할 때 발생하는 염증성 액체는 사혈(부항)로 제거하는 것이 도움이 됩니다.

It is helpful to remove the inflammatory fluid generated when bacteria die using bloodletting (cupping).

피부 가까운 곳에서 발생한 염증성 액체는 인체 내부에서 감당하기 어려울 만큼 많이 발생하기 때문입니다.

This is because the amount of inflammatory fluid generated near the skin is too much for the body to handle.

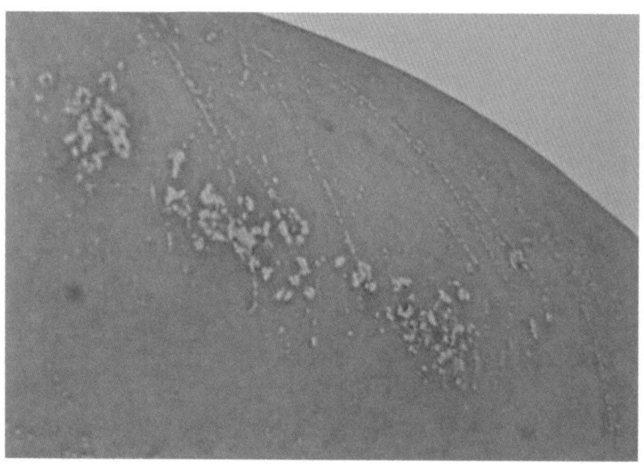

**소독 후 샤워를 마친 피부 상태(마이코플라즈마균의 이동 모습과 거의 동일)**

**Skin condition after showering after disinfection (almost identical to the movement of mycoplasma bacteria)**

## 네 번째 조치사항과 그 결과

## Fourth action and its results

### 1. 지금까지의 진행과정 평가

### 1. Evaluation of progress so far

1) 유기규소이온액 기반 소독용 화장품은 효과가 있었음. 그러나, 소독은 극심한 고통을 수반하였음.

1) Disinfecting cosmetics based on organosilicon ionic liquid were effective. However, sterilization was extremely painful.

2) 소독용 화장품을 발랐을 때, 통증을 수반하는 것으로 몸에 세포 내 기생충의 존재를 확인할 수 있음. (아토피 피부염의 완치 여부를 검증할 수 있는 방법이기도 함.)

2) When applying disinfectant cosmetics, the presence of intracellular parasites in the body can be confirmed by pain. (It is also a method to verify whether atopic dermatitis is cured.)

3) 유기규소이온액 기반 성장인자 화장품은 가려움증을 해소하는 데 효과가 있었음. (그러나 균이 많을 경우 세균의 군집생활 때문에, 즉 중증 아토피의 경우 가려움증을 해소하

는 데 어려움을 겪음.)

3) Organosilicon ionic liquid-based growth factor cosmetics were effective in relieving itching. (However, if there are a lot of bacteria, it is difficult to relieve itching in the case of severe atopy due to the bacterial colony life.)

4) 아토피균이 몰려 있는 중증 아토피의 경우 균이 소독에 저항하여 가려움증 해소에 어려움을 겪었음. (균이 많을 경우 항생제 등의 도움이 필요하다고 판단됨.)

4) In the case of severe atopy where atopic bacteria were concentrated, it was difficult to relieve itching because the bacteria resisted disinfection. (If there were a lot of bacteria, it was judged that help such as antibiotics was necessary.)

2. 항생제 복용결정: 중증 아토피에 준하는 상황이라고 판단하여 항생제(아지스로마이신)를 의사의 진단 아래 처방받아 복용키로 하였다. (52년간 비염을 가지고 있는 동안 균이 증식되었다고 판단)

2. Deciding to take antibiotics: It was determined that the condition was similar to severe atopic dermatitis, and the patient was prescribed antibiotics (azithromycin) under the doctor's diagnosis and decided to take the patient. (It was determined that bacteria had proliferated

during the 52 years of rhinitis)

1) 항생제 선택기준(Antibiotic selection criteria):

(1) 마이코플라즈마 폐렴의 경우 마이코플라즈마균에 대한 치료는 1차적으로 마크로라이드계 항생제를 투여합니다.
(마이크로라이드계 항생제의 종류: 에리스로마이신, 아지스로마이신, 클라리스로마이신)
(테트라사이클린계 항생제의 종류: 독시사이클린, 에라바사이클린, 미노사이클린, 오마다사이클린, 테트라사이클린)
(플루오로퀴놀론계 항생제 종류: 시프로플록사신, 델라플록사신, 제미플록사신, 레보플록사신, 목시프록사신, 노르플록사신, 오플록사신)

(1) In the case of mycoplasma pneumonia, the primary treatment for mycoplasma bacteria is macrolide antibiotics.
(Types of microlide antibiotics: erythromycin, azithromycin, clarithromycin)
(Types of tetracycline antibiotics: doxycycline, eravacycline, minocycline, omadacycline, tetracycline)
(Types of fluoroquinolone antibiotics: ciprofloxacin, delafloxacin, gemifloxacin, levofloxacin, moxiproxacin, norfloxacin, ofloxacin)

(2) 이중 아지스로마이신을 선택한 이유는 환자가 비염(아토피증후군)에서 발현된 아토피 피부염인 점을 고려하였습니다.

(2) The reason for choosing azithromycin was considering that the patient had atopic dermatitis caused by rhinitis (atopic syndrome).

(3) 또한, 아지스로마이신(화이자 상품명 지스로맥스 250mg)은 미국 FDA가 2004년경 세균성 부비동염(ABS)에 3일 요법으로 사용하도록 승인한 항생제임도 고려하였습니다.

(3) In addition, we also considered that azithromycin (Pfizer brand name Zithromax 250 mg) is an antibiotic approved by the US FDA for use as a 3-day regimen for bacterial sinusitis (ABS) around 2004.

(4) 아울러 투여일수는 아토피 피부염에 적용한 사례가 없기 때문에 500mg/day 15일 섭취분을 처방받아 경과를 보아가며 섭취하였습니다.

(4) In addition, since there have been no cases of application to atopic dermatitis regarding the number of days of administration, 500 mg/day (morning and evening) was prescribed for 15 days and was taken while observing the progress.

2) 소독용 화장품을 이용한 소독 주 1회(완치 여부 확인용)

2) Disinfection using disinfectant cosmetics once a week (To check complete cure)

3) 성장인자 화장품 매일 수시로 사용

3) Use growth factor cosmetics every day

### 3. 아지스로마이신 항생제 복용결과(Results of antibiotic (azithromycin) use):

**1일차**: 피부 상태와 사진에 찍힌 균 상태를 보여주고 항생제를 처방받음.

**1 Day**: Show your skin condition and the condition of the bacteria in the photo and receive a prescription for antibiotics.

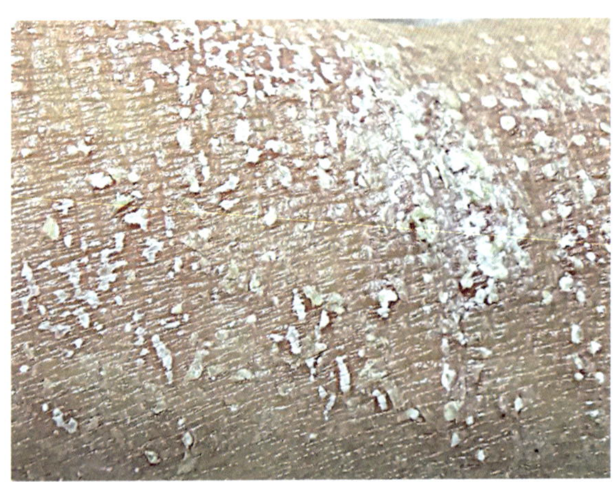

**피부로 튀어나왔던 균들의 모습**

**The appearance of bacteria protruding from the skin**

**2일차**: 따가움과 가려움이 심하게 나타남. '성장인자 화장품'을 사용해도 소용없었고, 과산화수소 소독도 소용없었음. 오직 뜨거운 물로 하는 샤워가 도움이 됨.

**2 Day**: Severe stinging and itching occurs. Using 'growth factor cosmetics' didn't help, and hydrogen peroxide disinfection didn't help either. Only hot showers help.

**3일차**: 따가움과 가려움이 더욱 심하게 나타남. 하루 종일 고통스러웠음.

**3 Day**: Stinging and itching become more severe. I was in pain all day.

**4일차**: 따끔거림과 가려움이 점차 개선이 되며, 아토피 피부염이 심한 곳에 딱딱한 각질이 형성되었음. 이때부터 성장인자 화장품을 바르면 가려움증이 해소되었음. (세균수가 많을 때는 가려움증이 통제되지 않다가, 세균수가 줄어들면서 가려움증이 통제되는 것으로 해석됨.)

**4 Day**: Stinging and itching gradually improved, and hard dead skin cells formed in areas with severe atopic dermatitis. From then on, the itching was relieved by applying growth factor cosmetics. (It is interpreted that when the number of bacteria is high, itching is not controlled, but as the number of bacteria decreases, itching becomes controlled.)

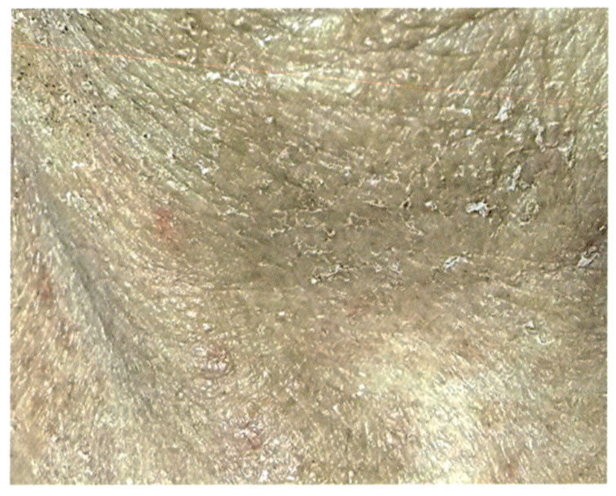

**항생제를 먹고 생긴 각질**

**Photo of dead skin cells that appear after taking antibiotics**

**5일차:** 약간 가려움이 올라왔으나, 따가움이 사라졌다. 목에 딱딱하게 올라왔던 각질이 '성장인자 화장품'을 발랐더니 부드러워졌다. 아토피가 심했던 팔도 따가움은 사려졌으며, '성장인자 화장품'을 사용하였더니 팔의 가려움증이 해소되었다. 몸속에 가려움증은 아직 남아있는데 전반적으로 피부가 깨끗해졌다.

**5 Day:** There was a slight itchiness, but the stinging disappeared. The hard dead skin cells on my neck became soft when I applied 'growth factor cosmetics.' The stinging sensation in my arm, which had severe atopic dermatitis, was relieved, and the itchiness in my arm was relieved when I used 'growth factor cosmetics.' The itchiness still remains, but overall my skin has cleared up.

**6일차**: 아토피 피부염이 가장 심했던 팔오금만 콕콕 찌르는 고통이 아직 남아있다. 피부 속으로 물집이 생겨서 긁어 터트렸는데 그 이후 시원해지면서 가려움증이 해소되었다. (다른 모든 피부에서는 찌르는 통증이 사라졌고, 가려움증은 '성장인자 화장품'을 바르면 즉시 해소되고 있다.)

**6 Day**: I still feel the pain of being pricked on my elbows, where the atopic dermatitis was most severe. A blister formed inside my skin, so I scratched it to burst it, but after that it cooled down and the itchiness went away. (For all other skin types, the stinging pain has disappeared, and the itchiness is immediately relieved by applying 'growth factor cosmetics.')

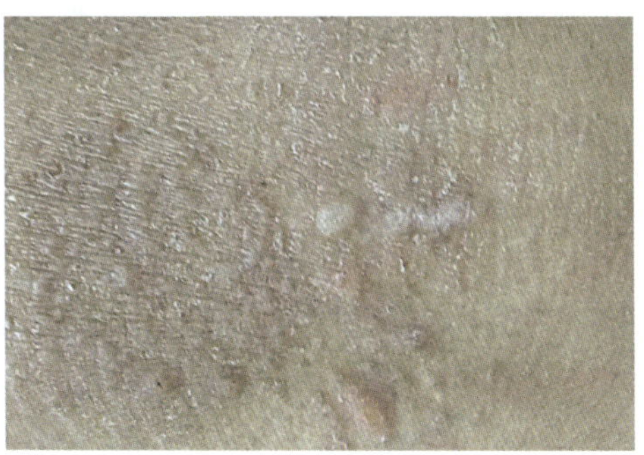

**아토피 피부염이 심했던 팔오금의 상태사진**

**Photo of an arm with severe atopic dermatitis**

피부 속에 형성된 물집이 보인다.

Blisters formed in the skin are visible.

**7일차**: 가려움은 발생하였는데 따가움은 발생하지 않았다. 항생제를 먹으면 따가움이 열 배로 올라왔었다. 그런데 오늘 아침부터 따가움이 전혀 올라오지 않고 있다. 가려움은 균에 의한 가려움이 아니라 딱지가 생기거나, 없어지면서 생기는 가려움 같다. 이 가려움은 아토피 피부염으로 인한 가려움과 다르게 미세한 가려움이다. 코는 아직까지 막히는 상태이다. 피부를 보면 피부 속으로 까맣게 보이는 모래알 같은 딱지가 보인다. (가렵지 않다.)

**7 Day**: Itching occurred, but no stinging occurred. When I took antibiotics, the stinging increased tenfold. But since this morning, there has been no stinging at all. The itchiness is not caused by bacteria, but is like an itch caused by the formation or disappearance of a scab. This itch is a subtle itch, unlike the itch caused by atopic dermatitis. My nose is still clogged. When you look at the skin, you can see black, sand-like scabs inside the skin. (It's not itchy.)

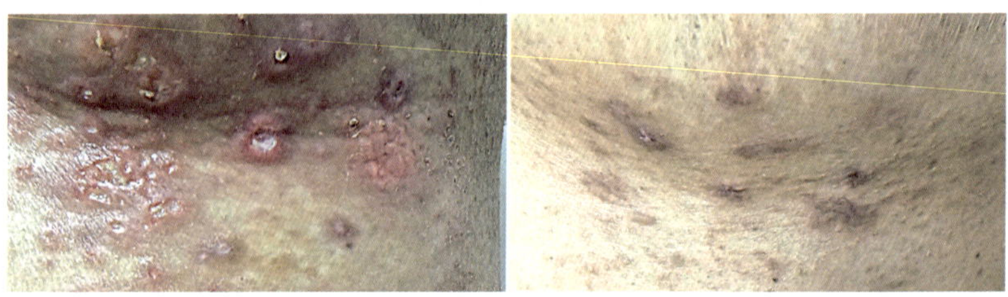

**아토피 피부염이 가장 심했던 상처 부위가 아물어가고 있다.**

**The wound area where atopic dermatitis was**

**most severe is healing.**

늦은 저녁 하체에 따끔거림이 약간 있었다. (몸 속에 아직 균이 남아있는 것으로 여겨졌다.)

Late in the evening, I felt a slight tingling sensation in my lower body. (It was believed that there were still bacteria remaining in the body.)

**8일차**: 온 몸에서 딱지가 나와서 6시간 이상 안티프라민으로 마사지를 하여 딱지를 제거했다.
**8 Day**: Scabs appeared all over my body, so I massaged with antipramine for more than 6 hours to remove the scabs.

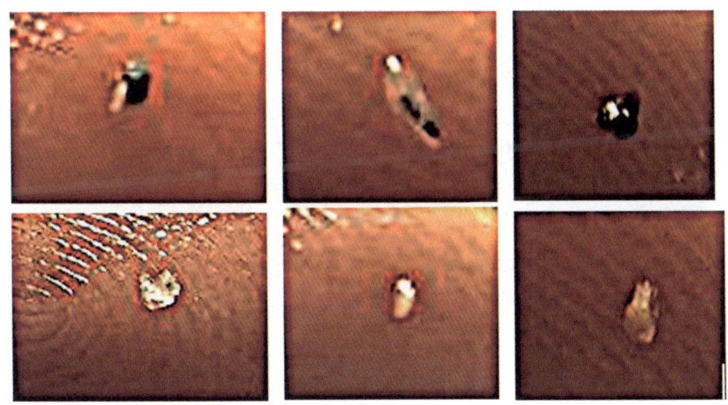

**다양한 모양과 색의 딱지들**

**Scabs of various shapes and colors**

가려움증과 따끔거림은 전반적으로 모두 멈춘 것으로 하루 종일 가려움증과 따끔거림을 느낄 수 없었다.

The itching and tingling stopped overall and I could not feel itching or tingling all day.

그러나, 피부 속에 모래알처럼 딱지가 생겼고, 피부 밑에 물집이 있는 것이 느껴짐.

However, I felt that scabs like grains of sand had formed in my skin and that there were blisters under my skin.

**9일차**: 새벽에 팔이 가려워서 성장인자 화장품으로 가려움증을 해소하였음. 몸에서 톡톡 쏘는 통증도 거의 없음.

**9 Day**: My arms were itchy in the morning, so I relieved the itchiness with growth factor cosmetics. There is almost no pain in the body.

**10일차**: 팔은 계속 성장인자 화장품을 덧발랐더니 안정을 찾았는데 주변 가슴하고 목 부분에 약간의 아토피 증세가 다시 올라왔음.

**Day 10**: I continued to apply growth factor cosmetics to my arms and they stabilized, but some symptoms of atopic dermatitis returned to my chest and neck.

**11일차**: 목, 어깨, 팔에 물집이 올라와서 터트렸음. 물집이 있다는 것은 균이 있어서 항생제에 저항하고 있다는 것으로 보임.

**Day 11**: Blisters appeared on my neck, shoulders, and arms and burst. The presence of blisters indicates that there are bacteria that are resistant to antibiotics.

**12일차:** 성장인자 화장품만 열심히 써도 균이 줄어드는 것이 느껴짐. 계속 바르면 아토피 피부염 부위가 줄어드는 것이 눈에 보임.
**Day 12:** Even if I diligently use growth factor cosmetics, I can feel the bacteria decreasing. If you continue to apply it, you will see that the area of atopic dermatitis is decreasing.

극도로 가려움이 시작되었는데 성장인자 화장품을 지속적으로 덧바르면서 딱지가 생기고 서서히 가려움증이 잡히기 시작했음.

I started to feel extreme itchiness, but as I continued to reapply growth factor cosmetics, scabs formed and the itchiness gradually subsided.

**13일차:** 피부 상태가 좋아졌으며 한쪽 팔에만 아토피 피부염이 몰려서 나타났음.
**Day 13:** Skin condition improved, and atopic dermatitis appeared only on one arm.

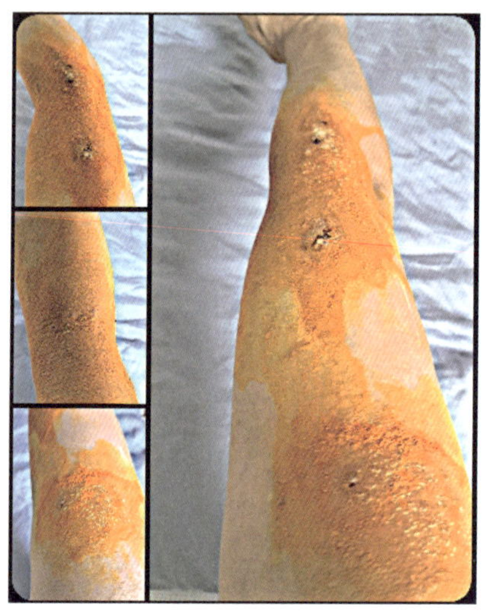

**3시간 소독하는 사진(이제는 소독용 화장품으로 3시간 이상 소독을 해도 예전처럼 심한 통증이 유발되지 않음. 그러나 심한 붓기가 발생함.)**

Photo of disinfection for 3 hours (Now, even if disinfected with disinfectant cosmetics for more than 3 hours, severe pain does not occur as before. However, severe swelling occurred.)

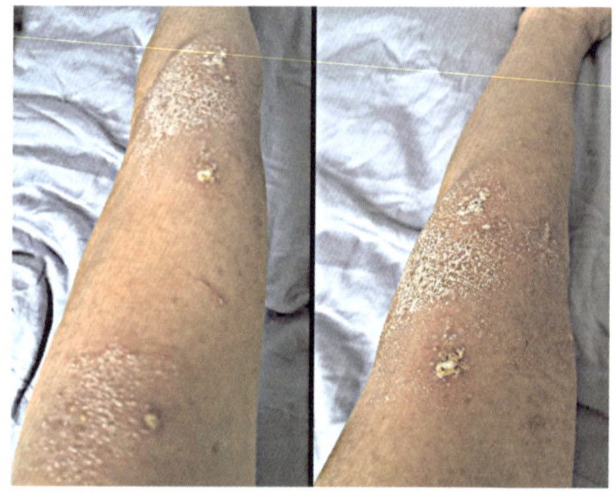

**3시간 소독 후 원인균이 피부에 올라온 모습(처음보다 균의 양이 현격하게 줄어들었다.)**
**After 3 hours of disinfection, the causative bacteria appear on the skin (the amount of bacteria has decreased significantly compared to the first time.)**

**14일차:** 전신 아토피 피부염은 전반적으로는 가라앉았는데 특이하게 한쪽 팔에만 균이 몰려 있는 것으로 보임. (팔의 아토피 피부염 부위는 코끼리 다리처럼 부어 올랐음.)
**Day 14:** As for systemic atopic dermatitis, the atopic dermatitis has subsided overall, but the bacteria appear to be concentrated only on one arm. (The atopic dermatitis area on the arm was swollen like an elephant's leg.)

**부어 있는 팔의 모습(아직 균이 보인다.)**
**Swollen arm (bacteria still visible)**

이제는 소독 시에 쑤시는 통증도 성장인자 화장품을 바르면 통증이 멈추기 시작했음.

Now, the aching pain during disinfection has started to stop when I apply growth factor cosmetics.

성장인자 화장품을 바르면, 쑤시는 통증과 가려움증이 통제되고, 딱지가 생김. (성장인자 화장품을 계속 덧바르고 있음.)

When growth factor cosmetics are applied, throbbing pain and itching are controlled, and scabs form. (I keep reapplying growth factor cosmetics.)

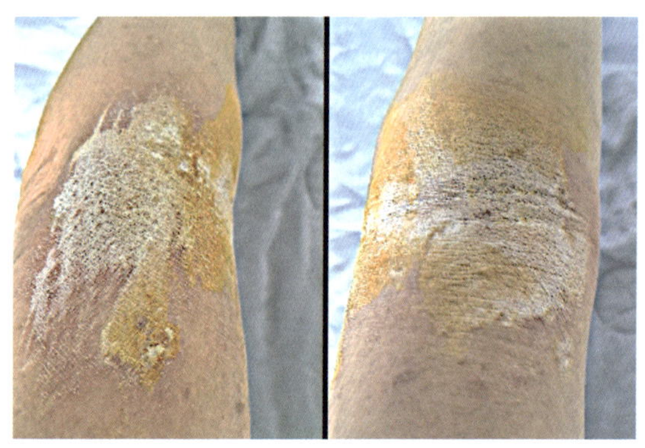

**장시간 소독 후 세포 모양으로 나타난 균의 모습**

**Appearance of bacteria in cell shape after long-term disinfection**

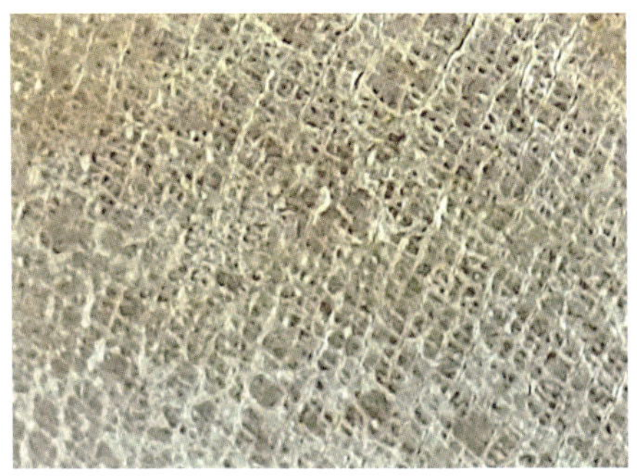

**나타난 균의 확대사진(마치 세포의 연결부위를 그대로 본뜬 것 같은 모습이다.)**

**An enlarged photo of the bacteria that appears (it looks like it mimics the connection area of cells.)**

성장인자 화장품을 덧바르는 이유는 피부에 산발적으로 아주 작지만 심한 가려움증이 올라와서 덧바르고 있는 것임.

The reason I reapply growth factor cosmetics is because I sporadically experience very small but severe itching on my skin.

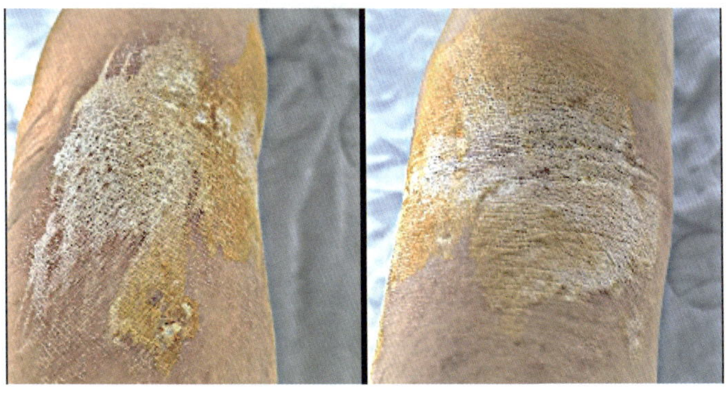

**팔 부위에 4시간 이상 소독용 화장품으로 소독한 모습(긴 시간 소독을 통해서 균이 남아있음을 확인했다.)**
**The arm area was disinfected with disinfectant cosmetics for more than 4 hours (it was confirmed that bacteria remained through long-term disinfection.)**

**15일차:** 한쪽 팔에만 딱딱한 게 존재함. 팔오금과 팔꿈치에만 아토피 원인균이 있는 것 같음. 전신이 아토피 피부염이 거의 없는데 한쪽 팔에만 남아있음.
**Day 15:** There is something hard on only one arm. It seems that only one arm has the bacteria that causes atopy. There is almost no atopic dermatitis all over the body, but it remains only on one arm.

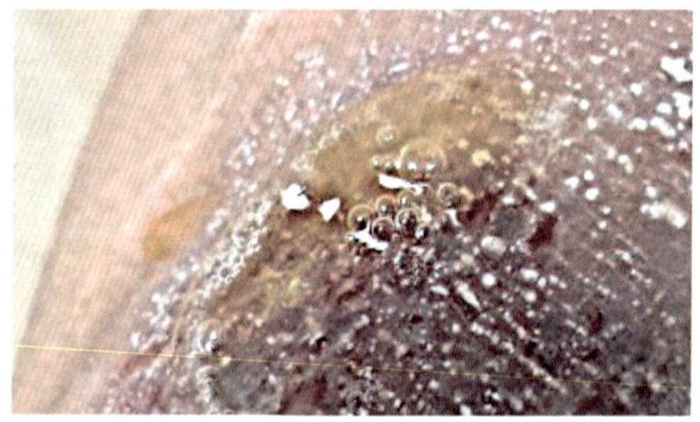

**부항침을 사용하지 않아도 염증성 액체가 빠져나온다. (이렇기 때문에 사혈을 통해 염증을 제거하는 것이 유리하다.)**
**Inflammatory fluid comes out even without cupping. (For this reason, it is advantageous to remove inflammation through bloodletting.)**

이 액체는 죽은 박테리아에 의해 발생한 염증성 액체인 것으로 관찰되었다.

The liquid was observed to be an inflammatory liquid caused by dead bacteria.

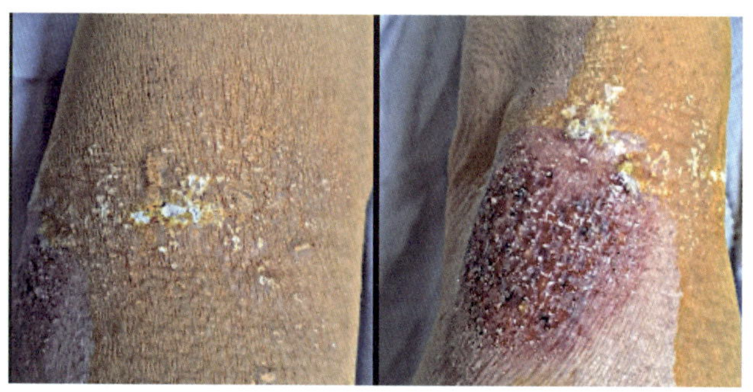

**한쪽 팔 심한 부위에 남은 아토피 원인균의 모습**

**Appearance of atopy-causing bacteria remaining in the severe area of one arm**

3시간 이상 소독에도 예전처럼 많은 균이 올라오지는 않는다.

Even after disinfection for more than 3 hours, there are not as many germs as before.

특이한 점은 성장인자 화장품을 바르면 가려움증이 완전히 조절되어 해소된다는 것이다.

What is unique is that when growth factor cosmetics are applied, the itching is completely controlled and resolved.

수시로 염증을 제거해주지 않으면 피부에 부종이 발생한다.

If inflammation is not removed regularly, swelling may occur on the skin.

## 다섯 번째 조치사항과 그 결과
Fifth action and its results

---

1. 항생제 아시트로마이신 섭취기록
   Antibiotic azithromycin intake record

   1) 섭취기간: 15일
   1) Intake period: 15 days

   2) 1일 섭취량: 250mg 2회 총 500mg
   2) Daily intake: 250mg twice, total 500mg

2. 아지스로마이신 섭취평가
   Azithromycin intake evaluation

   1) 비염은 조금 나아졌으나, 완치되지 아니함.
      Rhinitis improved slightly, but was not completely cured.

   2) 아토피 원인균은 15일 동안 확실하게 줄어든 것으로 평가됨.
      It was assessed that the bacteria causing atopy were significantly reduced over 15 days.

• 소독 시에 나오는 균의 양이 현저하게 줄어듦.
• The amount of bacteria released during disinfection is significantly reduced.

• 아토피 피부염이 중증으로 심할 경우 통제되지 않던 가려움증과 통증이 성장인자 화장품으로도 통제될 정도로 호전됨.
• When atopic dermatitis is severe, uncontrolled itching and pain improve to the point where they can be controlled with growth factor cosmetics.

• 확실히 아지스로마이신은 중증아토피 치료에 도움이 됨. (균을 줄이는 데에는 도움이 되나 완치되지 아니함.)
• Azithromycin is definitely helpful in treating severe atopic dermatitis. (It helps reduce bacteria, but does not cure it.)

**3) 결론: 15일 아지스로마이신 섭취에도 아토피 원인균이 박멸되지 아니함.**
**3) Conclusion: Even with azithromycin intake for 15 days, atopy-causing bacteria were not eradicated.**

### 3. 추가 조치사항: 독시사이클린 처방
### 3. Additional measures: Prescribe doxycycline

의사로부터 독시사이클린을 처방받음.

I was prescribed doxycycline by my doctor.

1) 1일 100mg 2회 복용(총 200mg 7일 복용)

1) Take 100mg twice a day (total 200mg for 7 days)

## 4. 독시사이클린 섭취기록

## 4. Doxycycline intake history

**1일차:** 독시사이클린 섭취 전부터 오른쪽 머리가 전체적으로 다 아프다. 오른쪽 머리를 포함해서 귀 속까지 다 아프다. 입술에 진물이 잡히더니 까맣게 변하더니 부분별로 벗겨지고 있다. (각질이 생긴 것으로 느껴진다.)

**Day 1:** Even before taking doxycycline, the entire right side of my head hurt. It hurts all over my ear, including the right side of my head. My lips were covered in ooze, turned black, and were peeling off part by part. (It feels like dead skin cells have formed.)

**2일차:** 가슴과 목 부위에 가려움이 올라왔고, 아토피 비부염이 심한 팔 부위에도 가려움이 올라왔다. 오른쪽 머리가 아픈 통증은 전반적으로 개선되었다. 독시사이클린을 먹어서 그런지 감당이 안 될 정도의 졸음이 몰려온다. 가려움증은 성장인자 화장품으로 쉽게 호전되었다.

**Day 2:** Itching occurred on the chest and neck, and also on the arm where atopic rhinitis was severe. The pain in the right side of my head has improved overall. Maybe it's because I took doxycycline, but I feel so drowsy that I can't handle it. The itching was easily improved with growth factor cosmetics.

**3일차**: 왼쪽 팔에만 아토피 증세가 심하게 남았고 피부 전반적으로 깨끗해졌음. 가려움증은 성장인자 화장품으로 통제가 되고 있고, 피부 전반적으로는 가려움증이 없음. 독시사이클린 섭취 후 심하게 졸리는 현상은 지속됨.

**Day 3**: Severe atopic dermatitis remained only on the left arm, and the skin overall became clearer. Itching is controlled with growth factor cosmetics, and there is no itching on the skin overall. Severe drowsiness persists after taking doxycycline.

**4일차**: 왼쪽 팔 심한 부위 주변에 염증성 액체가 가득한 것으로 느껴짐. 성장인자 화장품으로 가려움증은 쉽게 해소됨.

**Day 4**: The area around the severe area of my left arm felt like it was full of inflammatory fluid. Itching is easily relieved with growth factor cosmetics.

**5일차**: 목 뒷부분과 어깨 임파선 주변으로 사혈(부항)을 하여 염증성 액체를 제거하였음.

**Day 5**: Bloodletting (cupping) was performed on the back of the neck and around the shoulder lymph nodes to remove inflammatory fluid.

**6일차**: 전신에 가려움증이 98% 제거되었습니다. 딱지가 생기는 현상 때문에 생기는 가려움증만 조금 남아있는 상태입니다.

**Day 6**: 98% of itching throughout the body has been eliminated. Only a little bit of itching remains due to the formation of scabs.

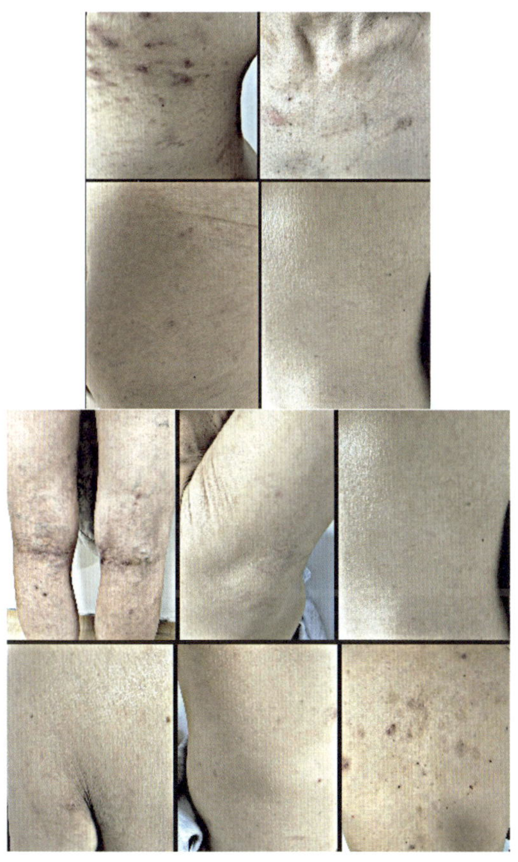

아토피 피부염의 흔적이 일부 남아있지만 전체적으로 염증은 모두 제거되었습니다.

Some traces of atopic dermatitis remain, but overall, all inflammation has been removed.

아토피 피부염 증세는 거의 모두 가라앉은 상태이나, 비염은 좋아지기는 하였으나 완치되지는 아니함.

Almost all atopic dermatitis symptoms have subsided, but rhinitis has improved but is not completely cured.

**7일차**: 더 이상 피부가 악화되지 않음. 이로써 항생제 아시트로마이신 15일과 독시사이클린 7일간 복용으로 아토피 피부염 치료에 긍정적인 효과를 볼 수 있었음.

**Day 7**: My skin is no longer getting worse. As a result, taking the antibiotic acithromycin for 15 days and doxycycline for 7 days showed a positive effect in treating atopic dermatitis.

내성균 발생을 억제하기 위해 독시사이클린을 4일간 추가적으로 복용할 예정임.

To suppress the development of resistant bacteria, doxycycline will be taken for an additional 4 days.

**그 외 추가사항(Additional information)**:

마이크로라이드계 항생제인 아지스로마이신과 테트라사이드계 항생제인 독시사이클린을 사용한 결과 아토피의 원인균 숫자를 줄이는 것은 확인되었으나, 테트라사이클린계보다 마이크로라이드계가 더 효과적으로 느껴진다는 것이 환자 의견이다.

**As a result of using azithromycin**:

A microlide antibiotic, and doxycycline, a tetraside antibiotic, it was confirmed that the number of bacteria causing atopic dermatitis was reduced, but the patient's opinion was that microlide antibiotics were more effective than tetracyclines.

특히 독시사이클린을 복용할 경우 속이 거북하거나, 졸리는 현상 등이 발생하였다.

In particular, when taking doxycycline, symptoms such as feeling uncomfortable or drowsy occurred.

비염(축농증) 치료에도 적당한 항생제가 치료에 효과적이나 완치를 위한 추가 연구가 필요함.

Appropriate antibiotics are effective in treating rhinitis (sinusitis), but additional research is needed to achieve a complete cure.

치료기간 중에는 먹는 유기규소이온액을 하루 30ml씩 매일 저녁에 섭취하였음. (향후 1년간 유기규소이온액을 섭취하면서 아토피 피부염의 재발 여부를 관찰할 예정임. 관찰결과 아토피 피부염이 재발하지 아니할 경우 완치로 판정.)

During the treatment period, 30 ml of organic silicon ion liquid was consumed every evening. (We plan to observe whether atopic dermatitis recurs while consuming organosilicon ion liquid for the next year. If atopic dermatitis does not recur as a result of observation, it is judged to be completely cured.)

이후 아지스로마이신 15일치를 추가 복용하고, 피부에 있는 아토피 원인균들을 99% 치료할 수 있었다. (1일 500mg 총 30일 복용)

Afterwards, I took an additional 15 days of azithromycin and was able to cure 99% of the bacteria that cause atopic dermatitis on the skin. (Take 500mg per day for a total of 30 days)

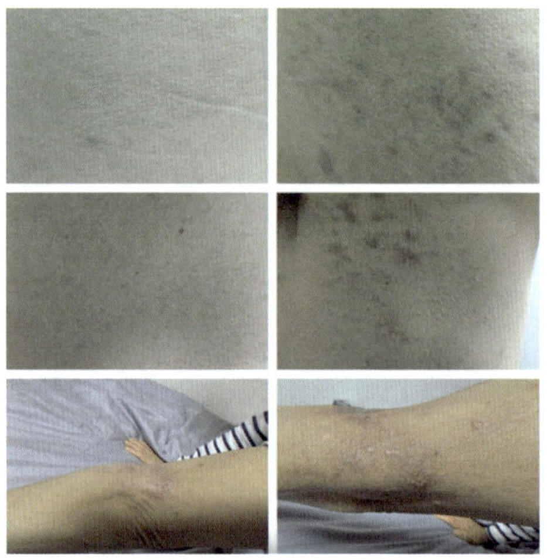

염증과 가려움증은 모두 제거되었으며, 아토피 피부염으로 인한 흉터는 서서히 없어지기 시작하였다. (이로써 치료를 중단하고 성장인자 화장품만 사용하고 있다.)

All inflammation and itching were eliminated, and the scars caused by atopic dermatitis gradually began to disappear. (With this, I stopped treatment and am only using growth factor cosmetics.)

# 부록 3 아토피 피부염 가려움증만 치료한 치료사례

# Appendix 3 Case study that only treated atopic dermatitis itching

"이 치료사례는 70대 남성의 아토피로 인한 전신 가려움증을 치료한 사례입니다."

"This treatment case treats whole body itching caused by atopy in a man in his 70s."

**사용물질**: 유기규소이온액 K9 500g + EGF 10ml + FGF 10ml + IGF 10ml (성장인자 화장품 제조)

**Materials used**: Organosilicon ionic liquid K9 500g + EGF 10ml + FGF 10ml + IGF 10ml (Growth Factor Cosmetics manufactured)

**사용방법**: 상기 사용물질만 1년에 걸쳐 피부 전체에 스킨 바르듯이 바름. (1일 1~3회)

**How to use**: Apply only the above ingredients to the entire skin over a period of one year. (1 to 3 times a day)

사용결과(Results of use):

1. 피부에 바르면 약간의 따가움을 느낀 후 곧바로 1분 이내에 가려움증이 해소됨.
1. When applied to the skin, a slight stinging sensation is felt and the itching is relieved within 1 minute.

2. 사용 초기에는 효과가 8시간 정도 지속되어 하루에 3번 정도 발라야 했음.
2. At first, the effect lasted for about 8 hours and was used about 3 times a day.

3. 3개월쯤 경과한 후 효과는 하루에 2번 정도만 발라도 될 정도로 지속되었음.
3. After about 3 months, the effect lasted long enough to require application only twice a day.

4. 9개월 경과한 시점에서부터 하루에 1번만 발라도 하루 종일 가려움증이 발생하지 아니하였음.
4. After 9 months, itching did not occur all day even when applied only once a day.

평가(evaluation):

환자의 연령이 70대이기 때문에 규소이온을 섭취할 경우 몸속에 있는 마이코플라즈마균이 얼마나 많이 피부 쪽으로 나와서 아토피 피부염을 발현시킬지 알 수 없어서 규소이온 섭취를 하지 않았을 경우 가려움증만 통제할 수 있는지 확인한 케이스임.

Because the patient was in his 70s, it was unknown how many mycoplasma bacteria in the body would come out to the skin and cause atopic dermatitis if silicon ions were ingested, so this was a case to see if only itching could be controlled if silicon ions were not ingested.

몸 속에 숨어있는 마이코플라즈마균을 치료할 수 있는 항생제 후보물질(아지트로마이신, 독시사이클린, 레보플록사신, 목시플록사신)이 아토피 치료에 도움이 되는지 의학계의 빠른 검증이 필요함.

The medical community needs to quickly verify whether antibiotic candidates (azithromycin, doxycycline, levofloxacin, and moxifloxacin) that can treat mycoplasma bacteria hidden in the body are helpful in treating atopy.

아토피의 원인이 세포 내 세균인 것으로 확인되었고, 세포 내로 물질을 전달할 수 있는 막수송체가 개발되었으므로, 다양한 형태의 아토피 치료제 연구가 지속되어야 함.

Since it has been confirmed that the cause of atopy is intracellular bacteria, and a membrane transporter capable of delivering substances into cells has been developed, research on various types of atopy treatments should continue.

## 부록 4 아토피 피부염 권장 치료방법에 관한 안내

## *Appendix 4* Information on recommended treatment methods for atopic dermatitis

이 안내는 병원 및 제약회사의 올바른 아토피 피부염 원인균에 대한 연구가 완료되어 치료방법이 확립될 때까지 사용하는 임시 방편에 불과합니다.

This guide is only a temporary measure to be used until hospitals and pharmaceutical companies complete research on the causative bacteria of atopic dermatitis and a treatment method is established.

사용 전에 의사의 진료를 통한 상담 후 적절한 치료방법을 선택하십시오.

Before use, consult a doctor and choose the appropriate treatment method.

### 1. 전체 치료대상자(아토피 원인균 보유 의심 대상)

### 1. All subjects eligible for treatment (subjects suspected of having atopy-causing bacteria)

아토피증후군 환자(아토피 피부염, 알레르기성 결막염, 알레르기성 비염(만성비염, 축농증 포함), 천식), 기관지염.

Patients with atopic syndrome (atopic dermatitis, allergic conjunctivitis, allergic rhinitis (including chronic rhinitis, sinusitis), asthma), bronchitis.

부인과에서 상재균으로 마이코플라즈마를 보균하고 있다고 진단받은 여성.

A woman diagnosed by a gynecologist as carrying Mycoplasma as a common bacterium.

폐렴, 피부선선, 류마티스 관절염, 다발싱경화증, 피부임 환자 등.

Patients with pneumonia, skin psoriasis, rheumatoid arthritis, multiple sclerosis, skin cancer, etc.

건강 유지의 목적으로 규소이온액을 섭취하였는데 피부에 가려움증이 유발되는 경우 포함.

This includes cases where silicon ion liquid is consumed for the purpose of maintaining health and itching occurs on the skin.

**2. 전체 치료대상자의 아토피균 보유 여부를 확인하는 방법**

**1) 보균자는 유기규소이온액 K8을 2개월 이상 6개월간 섭취하면서 피부 쪽으로 가려움증이 유발되는 현상을 보임. (권장기간 1년 이상 충분한 규소이온 섭취 권장 )**

Carriers show itching on the skin after consuming organosilicon ion liquid K8 for 2 to 6 months. (Sufficient silicon ion intake is recommended for a recommended period of at least 1 year)

2) 상재균이 많은 경우 소독용 화장품을 조제하여 전신에 도포하였을 경우 10분 이내에 균이 있는 부위에 통증이 유발됨. If there are a lot of bacteria, if you prepare disinfectant cosmetics and apply it to the whole body, pain will occur in the area where the bacteria are present within 10 minutes.

### 3. 중증 아토피 피부염 권장 치료방법에 관한 안내
### 3. Information on recommended treatment methods for severe atopic dermatitis

1) 본격적인 치료를 위하여 의사의 진료를 통한 적절한 항생제를 3주 내지 5주간 투여받아 균의 숫자를 줄이십시오. (잘 듣는 항생제일수록 심한 호전반응이 발생합니다.)
For full-scale treatment, consult a doctor and receive appropriate antibiotics for 3 to 5 weeks to reduce the number of bacteria. (The better the antibiotic, the more severe the improvement.)

(마이크로라이드계 항생제의 종류: 에리스로마이신, 아지스로마이신, 클라리스로마이신)

(Types of microlide antibiotics: erythromycin, azithromycin, clarithromycin)

(테트라사이클린계 항생제의 종류: 독시사이클린, 에라바사이클린, 미노사이클린, 오마다사이클린, 테트라사이클린)
(Types of tetracycline antibiotics: doxycycline, eravacycline, minocycline, omadacycline, tetracycline)

(플루오로퀴놀론계 항생제 종류: 시프로플록사신, 델라플록사신, 제미플록사신, 레보플록사신, 목시프록사신, 노르플록사신, 오플록사신)
(Types of fluoroquinolone antibiotics: ciprofloxacin, delafloxacin, gemifloxacin, levofloxacin, moxiproxacin, norfloxacin, ofloxacin)

• 본 책의 사례를 참고하여 적절한 항생제를 선택하시기 바랍니다.
• Please refer to the examples in this book to select an appropriate antibiotic.

• 아지스로마이신 1일 500mg × 15일 섭취 후 독시사이클린 1일 200mg × 14일(총 29일 섭취효과 있었음. 참고할 것)
• Azithromycin 500mg per day (250mg × 2 times a day) × 15 days followed by doxycycline 200mg per day (100mg × 2 times a day) × 14 days (Please note that intake was effective for a total of 29 days)

• 아지스로마이신을 1개월 내지 2개월간 꾸준히 섭취하는 것을 추천함(의사와 상담 후 처방받을 것)

• It is recommended to take azithromycin consistently for 1 to 2 months (obtain a prescription after consulting a doctor).

**2) 항생제를 투여하는 동안 경험해보지 못하는 피부의 변화를 견뎌낼 마음을 다지십시오.**

**2) Be prepared to endure changes in your skin that you may not experience while taking antibiotics.**

(1) 심한 가려움증, 콕콕 쑤시는 통증

(1) Severe itching and tingling pain

(2) 감당하기 힘든 각질 발생

(2) Generation of dead skin cells that are difficult to handle

(3) 외출을 하기 힘든 피부 상태

(3) Skin condition that makes it difficult to go out

**3) 항생제를 투여하는 기간을 포함하는 전 치료기간 동안 성장인자 화장품을 조제하여 수시로 몸 전체에 사용하십시오.**

**3) During the entire treatment period, including the period of antibiotic administration, prepare growth factor cosmetics and use them on the entire body at**

any time.

(1) 성장인자 화장품이 가려움을 통제하는지 관찰하십시오. (군집생활을 하는 균이 많을 경우 화장품을 이기고 더 가려움증이 극심해질 수 있습니다.)

(1) Observe whether growth factor cosmetics control itching. (If there are many bacteria living in colonies, they may overpower the cosmetics and cause more severe itching.)

(2) 소독용 화장품을 전신에 발랐을 때, 통증이 올라오는지 확인하십시오. (소독용 화장품은 균이 있을 때 그 숙주세포에만 반응합니다.)

(2) When applying disinfectant cosmetics to the entire body, check if pain occurs. (Disinfecting cosmetics only react to host cells when bacteria are present.)

(3) 치료기간 중 균이 많을 경우 군집생활을 하는 균의 특성상 성장인자 화장품이 가려움증을 해소할 수 없을 때가 있습니다. (이때는 소독용 화장품으로 소독을 하십시오.)

(3) If there are a lot of bacteria during the treatment period, growth factor cosmetics may not be able to relieve itching due to the nature of the bacteria living in colonies. (In this case, disinfect with

disinfectant cosmetics.)

(4) 항생제 및 소독으로 인해 균이 너무 많이 너무 빨리 죽어 염증이 피부 주변에 많을 경우 사혈(부황)을 사용하여 제거해주는 것이 도움이 됩니다.

(4) If too many bacteria are killed too quickly and there is a lot of inflammation around the skin due to antibiotics and disinfection, it is helpful to use bloodletting (cupping) to remove it.

(5) 피부에 아토피 피부염 증상뿐만 아니라 아토피로 인한 흉터도 모두 사라지는지 확인하십시오.

(5) Check whether not only the symptoms of atopic dermatitis on the skin but also the scars caused by atopic dermatitis disappear.

(6) 이 책에 나와 있는 사례집을 응용하십시오.

(6) Apply the case book in this book.

**4) 육안으로 아토피 피부염 치료가 끝난 것으로 확인되어도 1~2년간 유기규소이온액을 섭취하면서 가려움증이 다시 발생하는지 확인하십시오. (이 기간 동안에도 성장인자 화장품을 계속 사용하는 것을 권장합니다.)**

**4) Even if it is confirmed with the naked eye that the treatment of atopic dermatitis has been completed, check to see if itching occurs again**

while taking organosilicon ion liquid for 1 to 2 years. (It is recommended to continue using growth factor cosmetics during this period.)

5) 최종 치료가 완료된 후 의사의 진료를 통하여 완치판정을 받으십시오.
5) After the final treatment is completed, see a doctor and receive a diagnosis of complete recovery.

## 4. 경증 아토피 피부염 권장 치료방법 안내
## 4. Information on recommended treatment methods for mild atopic dermatitis

1) 경증환자의 경우 성장인자 화장품을 조제하여 꾸준히 발라서 피부염 상태가 호전되고 아토피로 인한 흉터가 사라지는지 확인.
1) For patients with mild symptoms, prepare growth factor cosmetics and apply it consistently to see if the dermatitis condition improves and scars caused by atopic dermatitis disappear.

2) 완전히 피부염 상태가 정상이 되고 가려움증이 올라오지 않으면, 규소이온의 섭취 여부를 결정하여 추가적으로 몸에 균이 있는지 확인할 것을 결정하십시오.
2) If the dermatitis condition becomes completely normal and itching does not occur, decide whether

to take silicon ions and additionally check for bacteria in the body.

3) 필요할 경우 의사와의 상담을 통해 적절한 항생제 치료를 병행하세요.
3) If necessary, consult with your doctor for appropriate antibiotic treatment.

## 5. 아토피로 인한 유기규소이온액 섭취에 대한 안내
## 5. Information on the intake of organosilicon ion liquid due to atopic dermatitis.

1) 유기규소이온액은 세포막수송체입니다.
1) Organosilicon ionic liquid is a cell membrane transporter.

2) 유기규소이온액의 섭취는 세포 속에 숨어있는 원인균을 피부 쪽으로 몰아내는 역할을 합니다.
2) Consumption of organosilicon ion liquid plays a role in driving out causative bacteria hidden in cells deep in the body towards the skin.

3) 아주 오랜 기간 몸 안에 아토피증후군 증상이나, 기타 세균성 피부염 증상 등을 가지고 있었던 환자의 경우 몸 안에 균이 얼마나 증식했는지 알 수 없음. (따라서 규소이온을 섭취하더라도 반드시 의사의 처방에 따라 충분한 항생제

섭취를 병행하여 치료하는 것이 필요함.)

3) In the case of patients who have had atopic dermatitis symptoms or other bacterial dermatitis symptoms for a very long period of time, it is unknown how much bacteria have proliferated in the body. (Therefore, even if silicon ions are ingested, it is necessary to receive treatment along with sufficient antibiotic intake according to the doctor's prescription.)

4) 나이가 너무 많아 균이 얼마나 나올지 모를 경우 가까운 병원을 방문하여 상담을 받아볼 것을 권장함. (나이 70 이상은 의사의 상담 후 치료 여부를 결정할 것. 치료의 효과도 중요하지만, 치료하는 과정에서 정신적 충격이 올 수도 있음.)

4) If you are too old and do not know how many bacteria you will have, it is recommended that you visit a nearby hospital and receive consultation. (If you are over 70, decide whether to receive treatment after consulting a doctor. Although the effect of treatment is important, mental shock may occur during the treatment process.)

5) 젊은 사람들은 유기규소이온액의 섭취와 아토피 치료를 병행할 것을 권장함.

5) It is recommended that young people combine

the intake of organosilicon ion liquid with atopic dermatitis treatment.

## 6. 소독용 화장품 제조방법

## 6. Method of manufacturing disinfectant cosmetics

1) 규소이온 기반 종합미네랄(한국 상품명 미네랄 밸런스) 30ml
1) Silicon ion-based comprehensive mineral (Korean brand name: Mineral Balance) 30ml

2) 화장품 원료 유황이온액 100ml
2) Cosmetic raw material sulfur ion solution 100ml

3) 구입처: www.si-ion.com
3) Where to purchase: www.si-ion.com

4) 유황이온액 30ml + 미네랄 발란스 30ml를 희석하여 소독용 화장품을 제조
4) Disinfecting cosmetics were prepared by diluting 30ml of sulfur ion solution + 30ml of mineral balance

## 7. 성장인자 화장품 제조방법

## 7. Method for manufacturing growth factor

cosmetics

| 투입순서<br>Input order | 원재료명<br>Raw material name | 소요량<br>Amount |
|---|---|---|
| 1 | 규소이온 K9<br>Chelated silicon ionic liquid K9 | 500.00g |
| 2 | EGF 1ppm(10ml) | 7.50g |
| 3 | FGF 1ppm(10ml) | 7.50g |
| 4 | IGF 1ppm(10ml) | 7.50g |
| 5 | MSM | 7.50g |
| 6 | 라우릴글루코사이드(유화제)<br>Lauryl Glucoside (emulsifier) | (약간) 0.25g |
| 7 | 아르기닌(pH 조절제)<br>L-arginine(pH adjuster) | 10.0g |
|  | 제조총량<br>Total manufacturing quantity | 585.6g |

# 부록 5 일반적인 유기규소이온 사용사례

# Appendix 5 Common Chelated Silicon Ion Use Cases

이 사례는 아토피 이외에 규소이온액의 사용 사례입니다.

This case is an example of the use of silicon ion liquid in addition to atopy.

규소를 포함한 모든 미네랄은 이온으로 섭취하는 것이 안전하며, 특히 킬레이트 방식의 이온섭취는 가장 안전하고, 흡수량이 많습니다.

It is safe to take all minerals, including silicon, as ions, and in particular, chelated ion intake is the safest and has the highest absorption rate.

유기규소이온액은 킬레이트 방식의 규소 미네랄 이온으로서 수용성 규소로 불리우는 콜로이드와 완전히 다른 물질임을 이해해야 합니다.

It must be understood that organosilicon ionic liquid is a chelating type of silicon mineral ion and is a completely different substance from the colloid called water-soluble silicon.

또한 메틸기($CH_3$)를 사용하는 비이온성 액체와도 다릅니다.

It is also different from nonionic liquids that use methyl groups ($CH_3$).

이 사용사례와 비슷한 사용자는 규소이온을 사용하신 후 반드시 병원 등 관계기관을 통하여 자신의 신체에 잘 적용되었는지 확인하시기 바랍니다.

Users similar to this use case should check with a hospital or other relevant institution to check whether the silicon ion has been properly applied to their body after using it.

이 사례는 개발자가 지금까지 사용자들과의 대화를 통해 있었던 일들을 모아놓은 것입니다.

This case is a compilation of what the developer has learned through conversations with users so far.

의학적으로 호전반응이 있는 효과들은 추후 유기규소이온액을 사용하더라도 본인의 몸이 좋아졌는지 병원의 진료로 확인하시기 바랍니다.

For medically beneficial effects, please check with your hospital to see if your body has improved even if you use organosilicon ion liquid in the future.

### 1. 소뇌위축증(Cerebellar atrophy) 호전사례 improvement

**case**

소뇌위축증(Cerebellar atrophy)은 굉장히 광범위한 개념이며, 일종의 질환군이라고 볼 수 있습니다. 대개 후천적인 혹은 2차적인 원인이 없이 서서히 소뇌에 퇴행성 변화가 오는 경우를 총칭합니다.

Cerebellar atrophy is a very broad concept and can be viewed as a group of diseases. It generally refers to cases where degenerative changes occur gradually in the cerebellum without any acquired or secondary causes.

규소이온을 섭취하신 환우가 약 6개월 섭취 경과를 말하였는데 소뇌위축증 진단을 받은 후 처음으로 병원에서 병의 진행이 중단되었다는 의사 소견을 받았다고 하였습니다.

The patient who consumed silicon ions said that it had been about 6 months since he had consumed them, and for the first time since being diagnosed with cerebellar atrophy, he received a doctor's opinion from the hospital that the progression of the disease had stopped.

질병 발생 전으로 완치된 것은 아니고, 질병의 진행이 중단되었다는 소견이었음을 참고하시기 바랍니다.

Please note that this does not mean that the patient has been completely cured, but that the progression of the disease has stopped.

## 2. 치매의 개선사례

## 2. Cases of improvement in dementia

치매의 경우 여러 사례가 존재합니다.

There are several cases of dementia.

유기규소이온액을 섭취하고 치매의 진행이 중단된 사례는 많습니다.

There are many cases where the progression of dementia was halted after taking organosilicon ion liquid.

치매환자의 가족이라면 6개월 단위로 유기규소이온액 섭취 후 병원에서 치매의 진행이 중단되었는지 확인하시기 바랍니다.

If you are a family member of a dementia patient, please check with the hospital to see if the progression of dementia has stopped after consuming organosilicon ion liquid every six months.

그러나, 유기규소이온액을 섭취하고 치매가 완치된 사례는 존재하지 않습니다.

However, there are no cases of dementia being completely cured by consuming organosilicon ion liquid.

규소이온이 치매를 예방하거나, 진행을 늦추거나, 진행을 중단하는 데에는 도움이 되는 것 같습니다.

Silicon ions appear to be helpful in preventing, slowing, or stopping dementia.

그러나, 치매가 발생하기 전으로 완치하는 것은 확인되지 않았습니다.

However, cure before dementia occurs has not been confirmed.

### 3. 파킨슨병의 사례

### 3. Case of Parkinson's disease

파킨슨병의 사례는 많지 않았습니다.

There were not many cases of Parkinson's disease.

사례가 많지 않은 이유는 규소이온을 섭취할 경우 파킨슨병 환자의 경우 극심한 통증에 시달리기 때문입니다.

The reason there are not many cases is because Parkinson's disease patients suffer extreme pain when silicon ions are ingested.

중증파킨슨 환자의 경우 유기규소이온액의 섭취를 매우 고통스러워하기에 도저히 섭취할 수 없다고 포기한 사례가 존재합니다.

In the case of patients with severe Parkinson's disease, there are cases where the ingestion of organosilicon ionic liquid was so painful that they gave up saying that it was impossible to ingest it.

30ml 한 번 섭취 후 섭취를 포기할 정도로 고통스러워했습니다.

After taking 30ml once, I was in so much pain that I gave up on taking it.

경증파킨슨 환자의 경우에도 섭취 시마다 미세한 고통을 호소하였습니다.

Even mild Parkinson's patients complained of slight pain every time they consumed it.

이러한 유기규소이온액을 섭취하면서 고통을 느끼는 사례는 파킨슨 질환이 유일하였습니다.

Parkinson's discase was the only case of feeling pain while consuming this organosilicon ion solution.

고통을 느끼는 이유를 유추하면 세포 속에 박혀 있는 알루미늄 미네랄과 치환반응에 의하여 자리바꿈을 하는 과정에서 일어나는 통증이라고 유추하였습니다.

The reason for feeling pain was inferred to be the pain that occurs in the process of switching places through a substitution reaction with aluminum minerals embedded in cells.

따라서, 중증파킨슨 환자나 경증이라 하더라도 파킨슨 환자는 병원의 도움을 받아 통증을 느끼지 않는 치료와 병행하지 않는 한 섭취가 매우 어려울 것으로 사료됩니다.

Therefore, it is believed that it will be very difficult for Parkinson's patients, even those with severe or mild Parkinson's disease, to consume unless they receive help from a hospital and receive pain-free treatment.

중증파킨슨 환자가 눈 앞에서 유기규소이온액 섭취를 포기할 정도의 고통을 느끼는 것을 보았고, 문제를 해결해줄 수 없었습니다.

I saw a patient with severe Parkinson's in such pain that he had to give up taking organosilicon ionic liquid right before my eyes, and I was unable to solve the problem.

파킨슨병은 걸리기 전에 예방이 중요하다는 생각을 했습니다.

I thought it was important to prevent Parkinson's disease before it occurs.

## 4. 불면증의 개선사례

## 4. Examples of improvement in insomnia

불면증에 대한 규소이온의 사례는 많은 편에 속합니다.

There are many examples of silicon ions for insomnia.

유기규소이온액을 2개월 동안 꾸준히 섭취한 결과 수면의 질이 좋아지는 것을 불면증 환자들이 느낍니다.

Insomnia patients notice that their sleep quality improves as a result of consistently consuming organosilicon ionic liquid for two months.

실제 유기규소이온액의 섭취시간을 저녁시간으로 고정한 이유가 잠이 많이 오기 때문입니다.

In fact, the reason why the intake time of organic silicon ion liquid is fixed to evening time is because it helps you sleep a lot.

15년 이상 병원치료에도 완치받지 못한 사람들조차도 2개월 경과하는 시점에서 불면증으로 인한 불편이 해소되었다고 합니다.

Even people who have not been cured despite hospital treatment for more than 15 years say that the inconvenience caused by insomnia has resolved after two months.

불면증 해소에 대한 불만사례는 없었습니다.

There were no complaints regarding insomnia relief.

가장 길게 불면증에서 벗어나지 못한 사례는 4개월이었습니다.

The longest case in which I was unable to get rid of insomnia was 4 months.

수면제를 사용하던 대부분의 사람들이 수면제를 끊었습니다. (4개월이 걸리지 않았습니다…)

Most people who used sleeping pills have stopped taking

them. (It didn't take 4 months…)

## 5. 당뇨병, 당뇨병성 족부질환

## 5. Cases of diabetes and diabetic foot disease

일명 당뇨발이라고 불리우는 발이 썩는 질환에 유기규소이온액을 먹고 환부에 발라준 결과 썩는 부위가 제거하지 않아도 될 정도로 재생되었습니다.

As a result of taking organosilicon ion liquid and applying it to the affected area for a disease called diabetic foot where the feet rot, the rotting area was regenerated to the point where it did not need to be removed.

또한 유기규소이온액은 고혈당 환자의 당 수치를 현격하게 낮추는 효과가 있습니다. (수치로 검증이 가능합니다.)

In addition, organosilicon ionic liquid has the effect of significantly lowering sugar levels in patients with hyperglycemia. (This can be verified numerically.)

거꾸로 저혈당 환자들은 유기규소이온액을 섭취할 때 주의하여야 합니다.

Conversely, patients with hypoglycemia should be careful when consuming organosilicon ion liquid.

## 6. 잇몸질환

## 6. Case of gum disease

잇몸이 약해져서 치아가 흔들리는 현상을 원상회복시켜 잇몸을 튼튼하게 합니다.

It strengthens gums by restoring weakened gums to their original state.

실제 치과병원에서 발치를 권유받았던 환자가 가글을 열심히 하고 유기규소이온액을 4개월 섭취한 후 다시 병원을 찾았을 때, 발치하지 않아도 될 것 같다는 의사의 소견을 들었습니다.

In fact, when a patient who had been recommended a tooth extraction at a dental hospital visited the hospital again after gargling diligently and taking organosilicon ion solution for 4 months, he heard the doctor's opinion that the tooth extraction would not be necessary.

잇몸을 튼튼하게 하여 치아를 견고하게 유지하게 합니다.

It strengthens the gums and keeps the teeth strong.

### 7. 혈관질환(뇌출혈, 뇌경색 포함)

### 7. Cases of vascular disease (including cerebral hemorrhage and cerebral infarction)

유기규소이온액은 막힌 혈관을 뚫어주고 혈관을 탄력 있고 튼튼하게 하여 혈관질환을 예방합니다.

Organosilicon ionic liquid unclogs blood vessels and prevents vascular diseases by making blood vessels elastic and strong.

또한 고혈압으로 인한 혈관 손상을 예방하는 데 도움이 될 수 있습니다.

It may also help prevent blood vessel damage in high blood pressure.

따라서, 유기규소이온액 섭취 시 호전반응으로 막힌 혈관 부위에 생기는 부종을 들을 수 있습니다. (약 10일 정도 부종이 발생하다가 해당 혈관이 뚫리면서 정상화됩니다.)

Therefore, when ingesting organosilicon ion liquid, you can hear edema occurring in blocked blood vessels as a positive reaction. (Swelling occurs for about 10 days and then normalizes as the blood vessels are opened.)

이러한 혈관을 뚫어주는 것 이외에 혈관의 탄력 유지에도 도움이 됩니다.

In addition to opening up these blood vessels, it also helps maintain the elasticity of blood vessels.

고혈압에 의한 뇌출혈, 또는 뇌경색 등을 예방하는 기능을 기대할 수 있습니다.

It can be expected to have the function of preventing cerebral hemorrhage or cerebral infarction caused by high blood pressure.

치매, 알츠하이머를 포함해서 뇌경색, 뇌출혈 등의 발생위험을 감소시키는 미네랄이라 할 수 있습니다.

It can be said to be a mineral that reduces the risk of dementia, Alzheimer's disease, cerebral infarction, and cerebral hemorrhage.

치료보다 예방에 초점을 맞추어야 할 듯합니다.

I think we should focus on prevention rather than treatment.

### 8. 얇은 손발톱 문제

### 8. Thin nail problem

손톱 발톱이 얇아서 찢어지거나, 깨지는 문제는 약 6개월 꾸준히 섭취 후부터 손발톱이 윤이 나고 단단해지는 것을 확인할 수 있습니다.

If you have a problem with your toenails being thin and tearing or breaking, you can see that your nails become shiny and hard after taking this product consistently for about 6 months.

### 9. 머리카락 염색 및 퍼머 시에 사용

### 9. Used for hair dyeing and perm

머리카락 염색 및 퍼머 시에 유기규소이온액의 일정량 첨가는 해당 약품에 의한 두피 부작용을 해소합니다.

Adding a certain amount of organosilicon ion liquid when dyeing or perming hair eliminates the scalp side effects caused by the drug.

뿐만 아니라 염색 및 퍼머 후 머리카락에 윤기를 더하여 줍니다.

In addition, it adds shine to hair after dyeing and perming.

**10. 화상 상처 치료**

**10. burn wound treatment**

화상을 입은 즉시 규소이온액을 환부에 바르면 빠르게 피부가 진정됩니다.

If you apply silicon ion liquid to the affected area immediately after a burn, it quickly soothes the skin.

화상으로 인한 쓰라림 등의 고통이 빠르게 진정됩니다.

Quickly soothes pain such as soreness caused by burns.

**11. 벌레에 물려 가려울 때**

**11. When you are bitten by an insect and itchy**

벌레에 물려 가려운 부위를 즉시 진정시킴.

Instantly soothes itchy areas bitten by insects.

24시간 내지 48시간 안에 상처 부위를 정상화시킴.

Normalizes the wound area within 24 to 48 hours.

# 부록 6 유기규소의 역사

# Appendix 3 History of organosilicon

## 1. 규소란 무엇인가?

## 1. What is silicon?

- 원소기호(element symbol): Si/Silicon(실리콘, 규소)
- 원자번호(atomic number): 14

- 자연 상태에서는 규소만 독립적으로 존재하지 않고 이산화규소($SiO_2$) 상태로 존재하며 규소의 함량이 100%이면 수정이 되고 다른 광물의 함유량에 따라 석영과 규석(차돌)이 됩니다.
- In the natural state, silicon does not exist independently, but exists in the form of silicon dioxide ($SiO_2$). If the silicon content is 100%, it becomes quartz, and depending on the content of other minerals, it becomes quartz or quartz stone (marble).

- 지구상에서 산소 다음 2번째로 많은 원소로 지구 전체 무게

의 약 18%를 차지합니다.

- It is the second most abundant element on Earth after oxygen, accounting for approximately 18% of the total weight of the Earth.

- 인체를 구성하는 가장 중요한 필수 미네랄입니다.
- It is the most important essential mineral that makes up the human body.

- 물에 거의 녹지 않아 식용으로 사용이 부적합하며 주로 반도체, 제철, 유리 가공 등 산업용으로 사용합니다.
- It is almost insoluble in water, making it unsuitable for human consumption. It is mainly used for industrial purposes such as semiconductors, steelmaking, and glass processing.

- "인체의 노화는 규소의 고갈로부터 시작된다."
- "The aging of the human body begins with the depletion of silicon."

인체의 60여 조개의 세포는 콜라겐이 접착제가 되어 각 세포를 강력하게 접착함으로써 인체의 형태가 유지되며 콜라겐을 만드는 유도체는 규소입니다.

Collagen acts as an adhesive for the 60 trillion cells in the human body and strongly adheres each cell to maintain the shape of the human body. The derivative that makes collagen is silicon.

- "골다공증은 칼슘 부족이라기보다 규소의 부족이다."
- "Osteoporosis is a deficiency of silicon rather than a deficiency of calcium."

규소는 체내에 흡수된 칼슘을 뼈로 운반하는 화물차와 같은 역할을 하며 뼈에 존재하면서 콜라겐을 만들어 규소가 운반해온 칼슘을 뼈에 부착시킵니다.

Silicon acts like a truck that transports calcium absorbed into the body to the bones, and while it exists in the bones, it creates collagen and attaches the calcium transported by silicon to the bones.

## 2. 유기규소의 역사
## 2. History of organosilicon

이 물질은 플래밍햄 코호트 연구와 관련이 있습니다.

This material is related to the Flamingham Cohort Study.

플래밍햄 코호트 연구(Framingham cohort study)는 1948년 메사추세스 플래밍햄 지방에서 30세에서 62세 사이의 건강한 남녀 5,209명으로 시작되었습니다.

The Framingham cohort study began in 1948 in Flamingham, Massachusetts, with 5,209 healthy men and women aged 30 to 62 years.

이후 지속적으로 이들을 2년마다 검진을 포함한 추적관찰을 시행하여 개개인의 생활습관, 유전요인 및 위험인자들이 특정 질병 발생에 미치는 영향에 관한 연구를 지속하여 왔습니다.

Since then, we have continued to conduct follow-up observations, including checkups every two years, and continue to conduct research on the impact of individual lifestyle habits, genetic factors, and risk factors on the occurrence of specific diseases.

플래밍햄 코호트 연구는 원래 20년간 예정으로 시작되었으나, 이후 여러 부분으로 나누어져 진행되었습니다. 1948년에 창설된 Orignal Cohort, 1971년 Offspring Cohort, 1994년 Omni Cohort, 2002년 Generation three Cohort 2003년 Omni Two Cohort 등이 만들어져 연구가 지속되고 있습니다.

The Flamingham Cohort Study was originally scheduled to last 20 years, but was later divided into several parts. Research continues through the creation of the Orignal Cohort in 1948, the Offspring Cohort in 1971, the Omni Cohort in 1994, the Generation Three Cohort in 2002, and the Omni Two Cohort in 2003.

우리나라에서도 몇몇 지역에서 시행되었으며, 강화 코호트, 원전 코호트, 충주 코호트 등이 이러한 연구에 해당됩니다.

It has been conducted in several regions in Korea, and these studies include the Ganghwa cohort, the nuclear power plant cohort, and the Chungju cohort.

플래밍햄 코호트의 대표적인 역학연구 결과 중 하나로서 규소(Silicon)의 인체에 미치는 영향이 일정 부분 규명되었다는 것이라 할 수 있습니다.

As one of the representative epidemiological research results of the Flamingham Cohort, it can be said that the effects of silicon on the human body have been identified to some extent.

**예를 들면(For example):**

- 규소가 칼슘보다 뼈를 더 튼튼하게 한다.
- Silicon strengthens bones more than calcium. (Source: https://www.ibric.org/myboard/read.php?Board=news&id=86516)

- 인체의 노화는 규소의 고갈로부터 시작된다. 규소는 콜라겐의 원료다.
- Aging of the human body begins with the depletion of silicon. Silicon is a raw material for collagen.

- 규소는 칼슘을 뼈로 운반하는 화물차 역할을 한다.
- Silicon acts as a cargo vehicle to transport calcium to bones.

등이 있습니다.

etc.

이 이외에 일본의 의사들은 다음과 같은 주장을 합니다:
In addition to this, Japanese doctors make the following claims:

- 규소는 인체의 모든 조직과 장기의 주요 재료가 되어 모든 질환의 치유에 절대적으로 필요하다.
- Silicon is the main material for all tissues and organs in the human body and is absolutely necessary for the healing of all diseases.

- 특히 암이나 난치성 질병의 원인이다.
- In particular, it is the cause of cancer and incurable diseases.

- 미토콘드리아 핵소체는 규소 덩어리다.
- The mitochondrial nucleolus is a lump of silicon.

이 연구결과가 말해주는 것은, 여러 가지가 있습니다만, 1940년부터 규소가 인체에 미치는 영향을 연구하였고, 그 연구결과가 발표되었다는 것입니다.
What this research result tells us is that, among many things, the effects of silicon on the human body have been studied since 1940, and the research results have been published.

그리고 이러한 연구는 가장 최근인 2012년 영국의 킬대학의 치매연구로 이어져 규소가 치매를 억제한다는 내용의 연구결과가 크리스토퍼 엑슬리 박사의 논문으로 알츠하이머병 저널(Journal of Alzheimer's Disease) 최신호에 실렸으며, 실험 참가자의 알루미늄 수치가 50~70% 감소한 것으로 확인되었습니다.

And this research was most recently followed by dementia research at Keele University in the UK in 2012, and the research results showing that silicon suppresses dementia were published in the latest issue of the Journal of Alzheimer's Disease in a paper by Dr. Christopher Exley. It was confirmed that the aluminum levels of the test participants decreased by 50-70%.

그리고, 이러한 영국의 연구는 국제특허로 이어졌습니다.

And, this British research led to an international patent.

이러한 규소에 대하여, 인간은 식물을 통한 섭취 이외에 플래밍햄 연구 이후에 지속적으로 규소를 직접적으로 섭취하기 위한 노력을 하였습니다.

Regarding silicon, humans have continued to make efforts to consume silicon directly since the Flamingham study, in addition to consuming it through plants.

그러나, 영국에서 출원한 이 특허는 이온을 만들지는 못했습니다.

However, this patent filed in the UK did not create an ion.

**〔유기 실리콘 역사의 시작〕**

[The beginning of the history of organic silicon]

유기 실리콘은 두 명의 프랑스 과학자인 Norbert Duffaut와 Lo c Le Ribault에 의해 발견되고 개발되었습니다.

Organic silicon was discovered and developed by two French scientists, Norbert Duffaut and Lo c Le Ribault.

Norbert Duffaut

Loic Le Ribault

1957년에 CNRS(National Center for Scientific Research)의 화학자이자 연구원인 Norbert Duffaut는 그의 작업에서 실라놀을 다루었습니다.

In 1957, Norbert Duffaut, a chemist and researcher at the National Center for Scientific Research (CNRS), addressed silanols in his work.

즉 인체에 쉽게 흡수될 수 있는 유기 실리콘 화합물의 형태를 연구하고 있었습니다.

In other words, we were researching a form of organic silicon compound that can be easily absorbed by the human body.

Norbert Duffaut의 연구결과가 발표된 후 유럽과 미국의 저명한 연구자들과 과학자들이 그의 연구를 계속했습니다.

After Norbert Duffaut's research results were published, prominent researchers and scientists in Europe and the United States continued his work.

1982년부터 1993년까지 Norbert Duffaut와 Lo c Le Ribault는 유기 실리콘의 새로운 분자개발에 협력했습니다.

From 1982 to 1993, Norbert Duffaut and Lo c Le Ribault collaborated on the development of new molecules of organosilicon.

Norbert Duffaut의 죽음 이후, Lo c Le Ribault는 두 작업을 계속했고 1994년에 메틸 실란 트리올이라는 분자에서 처음으로 구강소비용 유기 실리콘을 개발했습니다.

After Norbert Duffaut's death, Lo c Le Ribault continued both works and in 1994 developed the first organosilicone for oral consumption from a molecule called methylsilanetriol.

오늘날 실리카는 생체 이용률이 더 좋기 때문에 대신 사용됩니다. 30년 이상의 세심한 작업 끝에 마침내 유기 실리콘 생산 공식이 완성되었습니다.

Today, silica is used instead because it has better bioavailability. After more than 30 years of meticulous work, the organic silicon production formula was finally perfected.

이 두 사람에 의해 실라놀과 MMST라는 유기실리콘을 만날 수 있게 되었지만, 이 두 가지 물질도 결국 이온의 특성이 없는 분자체였습니다. (비이온성 액체입니다.)

These two people were able to discover organosilicon called silanol and MMST, but these two substances were ultimately molecular sieves without ionic properties. (It is a non-ionic liquid.)

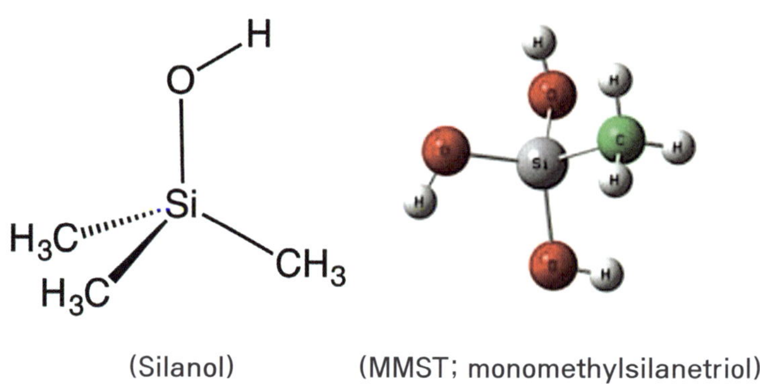


(Silanol) (MMST; monomethylsilanetriol)

2024년 올해로서 이런 규소를 먹을 수 있는 상태로 이온으로 만들고자 하는 노력은 67년이란 시간이 흘렀고, **한국에서 완전한 형태의 킬레이트 미네랄 이온으로 규소이온을 접하게 된 것**입니다.

As of this year, 2024, 67 years have passed since efforts were made to convert silicon into edible ions, and silicon ions have been introduced in Korea as a complete form of chelated mineral ion.

킬레이트 방식의 규소이온이 막수송체로서 작동한다는 사실은 세포 생물학 분야의 막수송체학 분야를 다시 기술해야 하는 놀라운 발견이 될 것입니다.

The discovery that chelating silicon ions function as membrane transporters will be a surprising discovery that will redefine the field of membrane transportomics in cell biology.

이로 인하여 세포에 물질 전달을 해야 하는 의학의 비약적인 발전과 화장품 분야의 비약적인 발전을 기대할 수 있습니다.

As a result, we can expect rapid advancements in medicine and the cosmetics field, which require the delivery of substances to cells.

## 3. 전 미국 생화학회장 로저 윌리엄스가 그의 저서에서 밝힌 규소의 효능

## 3. The efficacy of silicon revealed by Roger Williams, former president of the American Biochemical Society, in his book

### 1) 규소 결핍증

### 1) Silicon deficiency

- 족 관절 슬 관절 형성 불량
- Foot and knee joint dysplasia

- 관절 연골 중의 glycosaminoglycan(글리코사미노글리

칸) 함량 감소

- Decreased glycosaminoglycan content in articular cartilage

- 골다공증
- osteoporosis

- 결합조직의 탄성 결여
- Lack of elasticity of connective tissue

- 피부노화
- Skin aging

- 머리카락 손상
- Hair damage

- 손톱 손상
- Nail damage

### 2) 생리적 작용
### 2) Physiological action

- 콜라겐, 엘라스틴, 뮤코다당류, 히알루론산 등 탄소 골격에 결합되어 가교 형성에 관여한다.
- Collagen, elastion, mucopolysaccharides, hyaluronic acid, etc. are bound to the carbon skeleton and

participate in the formation of crosslinks.

- 결합조직의 구축이나 탄성에 관여한다.
- Involved in the construction and elasticity of connective tissue.

- 뼈의 석회화를 촉진한다. (뼈의 석회화가 왕성한 부위에 규소 함량이 많다.)
- Promotes bone calcification. (Silicon content is high in areas where bone calcification is active.)

- 콜라겐의 생성에 필수적이고 뼈 형성 초기과정에서 칼슘의 흡수를 촉진한다. (골다공증 예방)
- It is essential for the production of collagen and promotes calcium absorption in the early stages of bone formation. (prevention of osteoporosis)

- 손톱, 머리카락, 피부를 탄력성 있고 윤택하게 한다.
- Makes nails, hair, and skin elastic and shiny.

- 세 동맥의 탄력성을 유지한다. (심혈관계 질환 예방)
- Maintains the elasticity of the three arteries. (Prevention of cardiovascular disease)

- 알루미늄의 체내 축척을 방지한다. (알츠하이머병 예방)

- Prevents accumulation of aluminum in the body. (Alzheimer's disease prevention)

- 뼈와 연골을 포함한 다른 결합조직의 형성에서 기능하는 효소인 프로리하이드록실라제의 올바른 기능 유지에 필요하다.
- Necessary for maintaining the proper function of prolyhydroxylase, an enzyme that functions in the formation of bone and other connective tissues, including cartilage.

**3) 임상적 효능**

**3) Clinical efficacy**

- 골다공증
- Osteoporosis

- 골관절염
- Osteoarthritis

- 류머티즘성 관절염
- Rheumatoid arthritis

- 피부건강과 피부노화
- Skin health and skin aging

• 동맥경화증
• Arteriosclerosis

• 손톱, 머리카락 건강
• Nail and hair health

• 알츠하이머병 등의 예방과 치료 유지를 돕는다.
• Helps prevent and maintain treatment of Alzheimer' s disease, etc.

### 4) 참고사항(Note)

(1) 나이를 먹음에 따라 대동맥 중의 규소 함유량이 저하된다. 역학조사 연구에 의하면 동맥경화성 질환이 많은 문명국에서는 정제 식품을 섭취하기 때문에 규소 섭취량이 적다. 규소 함유량이 많은 섬유질 식품을 섭취하는 사람은 혈청(cholesterol)이 낮고 동맥경화도도 낮다.

(1) As we age, the silicon content in the aorta decreases. According to epidemiological studies, in civilized countries where arteriosclerotic disease is common, silicon intake is low because people consume refined foods. People who consume fiber foods high in silicon have lower serum cholesterol and lower arteriosclerosis.

(2) 규소는 자연계에서 원소로서 존재하지 못한다. 산화물(silica)로서 존재하며, 식물류 특히 섬유소에 아미노산과 결합된 이온으로 존재한다. 규소가 가장 많은 식품 중에는 alfalfa가 유용하다. (흡수율 양호)

(2) Silicon does not exist as an element in nature. It exists as oxide (silica) and as an ion bound to amino acids in plants, especially cellulose. Among the foods highest in silicon, alfalfa is useful. (Good absorption rate)

alfalfa는 calcium, magnesium, potassium, silicon 등이 매우 풍부하며, 소화기 궤양, 위염, 간장질환, 습진, 치질, 천식, 고혈압, 빈혈, 변비, 잇몸 출혈, 염증, 화상, 암, 부종 등에 자주 활용된다.

Alfalfa is very rich in calcium, magnesium, potassium, and silicon, and is often used for digestive ulcers, gastritis, liver disease, eczema, hemorrhoids, asthma, high blood pressure, anemia, constipation, gum bleeding, inflammation, burns, cancer, and edema.

(3) 규소의 보충은 피부 표면 바로 밑의 진피와 결합조직 지지구조의 두께를 증가시킨다.

(3) Supplementation of silicon increases the thickness of the dermis and connective tissue support structures just beneath the skin surface.

## 제1부를 마치며…
## Concluding Part 1…

유기규소이온액이 막수송체라는 사실은 기존 세포물리학 분야의 전문가들이 받아들이기는 쉽지 않을 것이다.

It will not be easy for experts in the field of cell physics to accept the fact that organosilicon ionic liquid is a membrane transporter.

그러나, 그동안 아토피 피부염 환자를 포함한 아토피증후군 환자들이 제대로 된 치료를 받지 못하였기 때문에 그 환자들을 치료하는 과정에서 너무나 쉽게 유기규소이온액이 세포막수송체라는 사실을 확인할 수 있을 것이다.

However, since atopic syndrome patients, including atopic dermatitis patients, have not received proper treatment, it will be easy to confirm that organosilicon ionic liquid is a cell membrane transporter in the process of treating these patients.

제대로 된 치료를 받지 못해 환자들이 넘쳐나고 있다.

Patients are overflowing because they are unable to receive proper treatment.

뿐만 아니라, 이 이온성 세포막수송체의 발견으로 세포생물학 분야의 발전과 함께, 약물 전달체계의 변화와 화장품 제조의 변화 등을 예상할 수 있다.

In addition, the discovery of this ionic cell membrane transporter can be expected to lead to developments in the field of cell biology, changes in drug delivery systems, and changes in the manufacturing of cosmetics.

그동안 세포 속에 숨어 살고 있던 세균 이외에 바이러스도 치료의 대상이 될 수 있을지도 모른다. (실제 경험 중에 쥐젖, 평편사마귀, 사마귀 등이 일부 제거된 사례가 있었는데 치료 전 증거를 확보해두지 못했기 때문에 거론하지 않았다.)

In addition to bacteria that have been living hidden in cells, viruses may also be targets of treatment. (During my actual experience, there was a case where some skin tags, flat warts, warts, etc. were removed, but I did not mention it because I could not secure evidence before treatment.)

이러한 사례들은 각 가정에서 알고 있는 것이 필요하기에 가정에서 관심 가지는 사람이 있어 꼭 읽어보기를 바란다.

These cases are necessary for each family to know, so if there is someone in your home who is interested, please be sure to read them.

이제 나머지 연구는 이 분야의 전문가의 몫이다. 필자는 물질 이온 분야를 연구하는 사람이지 의과학 분야에 문외한이다.

Now the rest of the research is up to experts in this field. The author is a researcher in the field of material ions and is unfamiliar with the field of medical science.

유기규소이온액을 개발하고, 우연히 유기규소이온액이 막수송체라는 사실을 알게 되고, 이로 인하여 아토피가 세포 내 기생충이 원인임을 알게 되었고, 부족하나마 치료방법을 알게 되었다.

Developing an organosilicon ionic liquid, By chance, I learned that organosilicon ionic liquid is a membrane transporter, As a result, it was discovered that atopy was caused by intracellular parasites. I finally found out how to treat it.

이 규소이온액이 세포막수송체라는 사실이 세포 생물학 분야에 어떤 영향을 끼치게 될른지 모르지만, 한국에 없는 과학 분야의 노벨상의 꿈을 이루어줄지도 모른다.

The fact that this silicon ion liquid is a cell membrane transporter. I don't know what impact it will have on the field of cell biology, but It may make the dream of a Nobel Prize in a field of science that does not exist in Korea come true.

그렇게 되길 희망한다.

I hope so.

# 기초물리학의 흠결

## Flaws in basic physics

인간은 에너지를 창조하는 종족으로 진화할 수 있는가?

Can humans evolve into a species that creates energy?

아니면 에너지를 소비하는 종족일 뿐인가?

Or are they just a species that consumes energy?

## 관념이 현실을 지배하는 세상
## A world where ideas dominate reality

우리는 '관념이 현실을 지배하는 세상에 살고 있다.'

We 'live in a world where ideas dominate reality.'

관념이란 무엇인가? 관념이란 '어떤 사물이나 견해에 대한 생각'이고 이러한 생각이 깊어져 우리가 알고 있는 과학적 상식이 된다.

What is an idea? An idea is 'a thought about an object or opinion,' and this thought deepens and becomes the scientific common sense we know.

예를 들어, 아인슈타인의 특수상대성이론이나 일반상대성이론, 뉴턴역학의 각종 법칙들이 관념에서부터 출발한 것이고, 이러한 관념이 세상을 현재까지도 지배하고 있는 것이다.

For example, Einstein's theory of special relativity, general relativity, and the various laws of Newtonian mechanics started from ideas, and these ideas still dominate the world to this day.

그러나, 아인슈타인은 '과학이 사회윤리적 목적을 이루는 도구를 제시할 뿐이며, 또한 전문가만이 그런 생각을 하는 것은 아니라고 했다.'

However, Einstein said, 'Science only presents tools to achieve social and ethical goals, and experts are not the only ones who think like that.'

하지만, 사실 전문가들은 아인슈타인과 뉴턴 등 한 시대를 풍미한 과학자들의 관념과 싸울 용기를 쉽게 내지 못한다.

However, in fact, experts do not easily muster the courage to fight against the ideas of scientists who dominated the era, such as Einstein and Newton.

왜 그럴까? 그것은 그들(아인슈타인, 뉴턴)이 만들어 놓은 관념의 토대 위에서 나 자신의 삶을 영위하고 있기 때문이다.

Why? This is because I lead my own life based on the ideas they (Einstein, Newton) created.

만일 이 시대를 살면서, 아인슈타인의 이론 중 어떠한 것이 틀렸다는 주장을 한다거나, 혹은 뉴턴역학의 법칙이 문제가 있다는 발언을 하는 물리학 교수나 물리학 박사가 있다면 화제의 중심에 설 것이다.

Living in this era, if there is a physics professor or doctor of physics who claims that any of Einstein's theories are wrong or that Newton's laws of mechanics are problematic, they will be at the center of discussion.

그것은 그 사람의 커리어에 매우 위험한 행위이다.

It is a very dangerous act for that person's career.

그는 위대한 과학자의 반열에 올라설 수 있을지도 모르지만, 실패하여 그의 이야기가 허무맹랑한 것으로 결론이 난다면 비난을 면치 못할 뿐만 아니라 지금까지 쌓아왔던 본인의 명성을 내동댕이치는 결과가 기다릴 것이기 때문이다.

He may be able to rise to the ranks of great scientists, but if he fails and his story is concluded to be empty, not only will he not be able to avoid criticism, but he will also face the consequences of destroying the reputation he has built up until now.

그러한 용기를 냈다는 것에 박수를 쳐야 하지만, 사람들은 박수보다 비난에 반응하기 쉽다.

You should applaud them for showing such courage, but people are more likely to respond to criticism than to applause.

또한, 과거의 그러한 과학적 법칙들이 지난 오랜 시간 동안 인간이 살아가면서 사용하는 데 커다란 문제가 없었다고 인식되는 상황에서 그러한 법칙의 오류를 주장하기란 더더욱 힘든 일이다.

In addition, it is even more difficult to assert the error of such scientific laws in the past when it is recognized that there have been no major problems in using them in human life for a long time.

계란으로 바위치기 같은 생각이 들지도 모른다.
It may feel like hitting a rock with an egg.

그래서 필자가 이 글을 쓰기 시작한 것이다.
That's why I started writing this article.

필자는 내동댕이칠 명성도 없고, 쌓아놓은 사회적 지위도 거의 없다.
I have no reputation to throw down and little social standing to build up.

필자는 이 글을 쓰기 시작할 무렵 특허심판원으로부터 심결문을 한 부 송달받았다.
Around the time I started writing this article, I received a copy of the trial decision from the Intellectual Property Trial and Appeal Board.

"금속 모재에 물질변환층과 침탄질화층을 형성하는 방법"에 관한 심결문이었다.
It was a trial decision regarding the patent "Method of forming a material conversion layer and a carbonitriding layer on a metal base material."

이 특허심판원의 심결문이 필자를 가슴 아프게 했던 것은 특허심판원이 개발자의 실험 오류를 의심했기 때문이다.
The reason the Patent Trial and Appeal Board's decision

made me heartbroken was because the Patent Trial and Appeal Board suspected an error in the developer's experiment.

**"물질 변환이 이루어진 것이라고 하는 것은 실험 오류가 포함될 수 있는 결과를 단순 제시한 것에 불과하고…"**

**"Saying that material conversion has occurred is merely a simple presentation of results that may contain experimental errors…"**

만일 필자가 아주 유명한 물리학자이자 논문을 수십 편을 발표하고, 사이언스나, 네이쳐지에 떠들썩하게 알려진 사람이라면 감히 개발자의 실험 오류를 들먹이지는 않았을 것이다. (적어도 실험을 해보기 전까지는…)

If the author were a very famous physicist who had published dozens of papers and was widely known in Science and Nature journals, I would not have dared to bring up the developer's experiment errors. (At least until I try the experiment…)

한국에서는 그러한 사이언스나, 네이쳐지에 논문이 실리는 것과 같은, 혹은 동일한 수준의 검증시스템을 일반인들이 이용할 수 있는 방법이 없다.

In Korea, there is no way for the general public to use the same level of verification system as those published in Science or Nature journals.

그러므로 앞으로도 나는 발명을 할 때마다 나의 실험 오류를 의심받을 것이라 생각한다.

Therefore, I believe that in the future, whenever I invent something, I will be questioned about my experimental errors.

나는 따라서 이 특허에 대한 특허심판원의 심결문에 대하여 특허법원에 소송을 제기하지 않았다.

Therefore, I did not file a lawsuit in the Patent Court against the Intellectual Property Trial and Appeal Board's decision regarding this patent.

특허 자체가 틀린 것은 아니다.

It is not that the patent itself is a mistake.

개발자의 실험이 의심받는 특허청에 아무리 내가 올바른 실험을 했다고 주장해도 그 주장이 받아들여질 가능성이 없는 것이 그 첫째 이유다.

The first reason is that no matter how much I claim that I conducted the correct experiment to the Patent Office where the developer's experiment is suspected, there is no chance of my claim being accepted.

그리고, 대한민국에서 상온핵융합이니, 물질 변환이니 하는 특허가 설사 등록된다 하더라도, 그것을 실용화하기 위한 기반이 갖추어져 있지도 않을 뿐더러 덩치가 너무 큰 일이기에 개인이 나설 문제가 되지 못하는 것이 둘째 이유다.

And, the second reason is that even if a patent for cold fusion or material conversion is registered in Korea, not only is there no foundation for commercializing it, but it is too large of a task to be a problem for individuals to step forward.

그 특허들은 그 당시에 그것을 내가 발명한 것으로 만족할 수 밖에 없기에 그냥 특허청에 공개되어 있기만 하면 될 일이라 생각했다.

Since I had no choice but to be satisfied with the patents as having been invented by me at the time, I thought it would be enough as long as they were disclosed to the patent office.

싸울 필요도 없고, 엄두도 나지 않았다.

There was no need to fight, and I didn't even dare.

그러나, 이 글은 조금 다르다.

However, this article is a little different.

이 글이 발표되면, 물리학계나 일반인들이 어떤 반응을 보일지 궁금하다.

I wonder how the physics community and the general public will react when this article is published.

다만, 기존의 학계나 일반인들이 특허청보다는 조금 나을 수도 있겠다는 생각을 해본다.

However, I think that existing academics and the general

public may be a little better than the Korean Intellectual Property Office.

이 글을 통해 과학을 있는 그대로 바라볼 자신도 있고, 이 글을 통해 관념이 지배하는 세상에서 그릇된 관념을 솎아낼 자신도 있다.

Through this article, I have the confidence to look at science as it is, and through this article, I have the confidence to weed out false ideas in a world dominated by ideas.

이제 이 글을 읽는 독자들이…

Now, readers of this article…

끝까지 글 내용이 재미 없지 않기를 바랄 뿐이다.

I just hope that the content of the article will not be boring until the end.

# 뉴턴과 아인슈타인
## Newton and Einstein

자신이 태어나기 152년 전에 사망한 아이작 뉴턴의 가속도의 법칙이 특수상대성 이론을 개발한 아인슈타인의 눈에는 어떻게 보였을까?

What did Isaac Newton's law of acceleration, who died 152 years before he was born, look like to Einstein, who developed the theory of special relativity?

아인슈타인의 눈에는 가속도의 법칙 공식이 자신이 만든 특수상대성이론의 공식과 매우 불일치한다고 생각했던 것은 아닐까?

In Einstein's eyes, did he think that the formula for the law of acceleration was very inconsistent with the formula for the special theory of relativity that he created?

가속도의 법칙에 의하면, 힘만 주어진다면, 기어이 광속을 돌파해버릴 것 같았기 때문에 이 가속도의 법칙을 고쳐야 한다고 생각했던 것 같다.

According to the law of acceleration, it seemed like if only force were given, it would easily break the speed of light, so it seemed like it was thought that this law of acceleration had to be changed.

이 아래 4개의 그래프는 서로 연관성이 있는 공식과 그래프들이다.

The four graphs below are formulas and graphs that are related to each other.

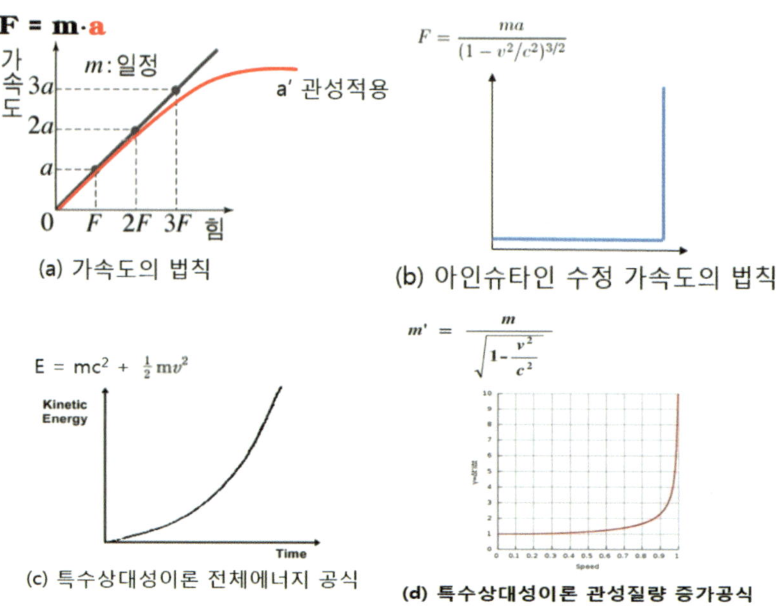

(a) 가속도의 법칙
(b) 아인슈타인 수정 가속도의 법칙
(c) 특수상대성이론 전체에너지 공식
(d) 특수상대성이론 관성질량 증가공식

(a) Law of acceleration (a' inertia application graph)

(b) Einstein's modified law of acceleration

(c) Special relativity overall energy formula

(d) Special relativity inertial mass increase formula

그중에 3개는 모두 아인슈타인에 의해 주장된 것이거나 수정된 것이다.

Among them, three are all claims made or modified by Einstein.

(a)는 가속도의 법칙에 의한 그래프와 해당 그래프에 관성의 법칙으로 인한 관성질량 증가현상이 적용되었을 경우 나타나는 그래프를 a'로 표현하였다.

In (a), the graph based on the law of acceleration and the graph that appears when the inertial mass increase phenomenon due to the law of inertia is applied to the graph are expressed as a'.

가속도의 법칙에서는 관성질량 증가현상이 적용되지 않았을 경우 정비례를 하지만 관성질량 증가현상이 적용되면, 해당 그래프의 모양이 달라진다.

The law of acceleration is directly proportional when the inertial mass increase phenomenon is not applied, but when the inertial mass increase phenomenon is applied, the shape of the graph changes.

관성의 법칙에 의한 관성질량 증가현상이 적용된 가속도의 법칙 그래프는 이렇게 그려졌다.

The acceleration law graph to which the inertial mass increase phenomenon due to the law of inertia is applied is drawn like this.

처음 F의 힘이 가해졌을 경우 가속도는 a가 된다.(ma) 이때 m에는 m'1의 관성이 누적된다.

When a force of F is first applied, the acceleration becomes a. (ma) At this time, an inertia of m'1 is accumulated in m.

여기서는 질량으로 m이 그리고 첫 번째 힘으로 인하여 생긴 m'1이라는 관성이 누적된 것이다. (가속도 a가 생겼을 때, 관성질점 m'1이 같이 생긴다.)

Here, m is the mass and the inertia m'1 created by the first force is accumulated. (When acceleration a occurs, inertial point m'1 also occurs.)

두 번째 2F의 힘이 가해졌을 경우 질량은 m + m'1가 되고 이 m'1는 첫 번째 F의 힘을 가해 만들어진 관성력(관성질량)이다.

When the second force of 2F is applied, the mass becomes m + m'1, and this m'1 is the inertial force (inertial mass) created by applying the first force of F.

이 m'1는 m과 함께 2F의 힘 중에서 일정량을 소비한다.

This m'1 consumes a certain amount of the force of 2F together with m.

따라서 2F의 힘이 가해졌을 때 m'1이 소비한 힘 때문에 F의 힘에서처럼 직선의 그래프가 나타나지 못하고, 기울기를 가지게 된다.

Therefore, when a force of 2F is applied, we have a crowd where the graph of Beautiful becomes smaller at a force of F due to the force expended by m'1.

그래서 결국 가속도는 2a 도달하지 못하게 된다.
So in the end, the acceleration does not reach 2a.

그리고 이때에도 2F의 가해진 힘에 의하여 m'2의 관성이 생긴다.
And even at this time, an inertia of m'2 is created due to the applied force of 2F.

또한 질량으로 m 질량 역할을 하는 m'1과 이 두 질량과 질량 역할을 하는 존재가 2F의 힘을 소모해서 m은 m'2의 관성을, m'1는 m'1-1의 관성을 만들어낸다.
In addition, m'1, which plays the role of mass m, and these two masses and the being that plays the role of mass consume 2F force, so that m creates the inertia of m'2 and m'1 creates the inertia of m'1-1. Pay it out.

다음 3F의 힘이 가해졌을 경우 질량은 m + m'1 + m'1 – 1 + m'2가 되고, 이 질량과 누적된 관성력들은 가해진 3F의 힘 중에서 일정량을 소비한다.
When the next 3F force is applied, the mass becomes m + m'1 + m'1 – 1 + m'2, and this mass and accumulated inertial forces consume a certain amount of the applied 3F force.

그리고 그 결과물로 m은 m'3를 m'1는 m'1-2를 m'1-1은 m'1-1-1'을 m'2는 m'2-1을 만들어낸다.

And as a result, m produces m'3, m'1 produces m'1-2, m'1-1 produces m'1-1-1', and m'2 produces m'2-1.

따라서 3F의 힘이 가해졌을 때 그래프는 기울기를 가지게 되며, 결국 가속도는 3a에 도달하지 못한다.

Therefore, when a force of 3F is applied, the graph has a slope, and ultimately the acceleration does not reach 3a.

즉, 누적된 관성은 완벽한 질량 역할을 한다.

In other words, the accumulated inertia acts as a perfect mass.

따라서 가속도 법칙에 관성의 법칙에 의한 관성질량 증가현상을 적용한 그래프 a'는 기울기를 가진 형태로 표현되며, 따라서 원래의 가속도 법칙의 그래프는 관성의 법칙이 적용되지 아니한 것이 명백하다.

Therefore, the graph a', which applies the phenomenon of inertial mass increase according to the law of inertia to the law of acceleration, is expressed in a form with a slope, and therefore, it is clear that the law of inertia is not applied to the graph of the original law of acceleration.

이 관성의 누적은 질량을 가진 물체가 광속을 벗어나기 어렵다는 것을 자연스럽게 표현하고 있다.

This accumulation of inertia naturally expresses the difficulty for objects with mass to escape the speed of light.

(b) 공식은 아인슈타인이 수정했다고 알려진 뉴턴역학의 가속도의 법칙 공식의 수정공식이다. 다시 공식을 살펴보자.

(b) The formula is a modified formula of the acceleration law formula of Newtonian mechanics, which is known to have been modified by Einstein. Let's look at the formula again.

$$F = \frac{ma}{(1 - v^2/c^2)^{3/2}}$$

이 공식을 가지고 시속 0km/s에서부터 광속(299,792,458km/s)에 이르기까지 그래프를 그려보면, "ᅩ" 모양의 그래프가 그려진다.

If you draw a graph using this formula from 0 km/s to the speed of light (299,792,458 km/s), a "ᅩ" shaped graph is drawn.

이것은 무엇을 뜻하는 것일까?

What does this mean?

(a) 그래프에서 관성의 법칙이 가속도의 법칙에 반영되지 않았다는 것을 설명하고, 관성의 법칙에 의한 관성질량 증가현상이 적용된 a' 그래프를 도출해냈다.

(a) In the graph, we explained that the law of inertia was not reflected in the law of acceleration, and derived the a' graph to which the phenomenon of inertial mass increase due to the law of inertia was applied.

그런데 (b)의 그래프는 관성의 법칙에 의한 관성질량 증가현상이 적용된 그래프와 전혀 상관없는 방향으로 전개되었을 뿐만 아니라, 원래 가속도의 법칙과도 상관없는 방향으로 전개되어 있다.

However, the graph in (b) not only developed in a direction completely unrelated to the graph to which the inertial mass increase phenomenon due to the law of inertia was applied, but also developed in a direction unrelated to the original law of acceleration.

즉, 아인슈타인이 수정했다는 수정된 가속도의 법칙 공식은 관성의 법칙을 위반하고 있고, 가속도의 법칙도 위반하고 있다.

In other words, the modified acceleration law formula that Einstein modified violates the law of inertia and also violates the law of acceleration.

이 말은, $(1 - v^2/c^2)^{(3/2)}$는 관성의 법칙도 무시하고, 가속도의 법칙 그래프도 무시하고 무조건 질량을 가진 물체는 광속을 벗어날 수 없다는 것을 표현하였을 뿐, 아무런 근거도 없는 공식이다.

This means that $(1 - v^2/c^2)^{(3/2)}$ ignores the law of inertia and the graph of the law of acceleration, and only expresses that an object with mass cannot escape the speed of light, and there is no basis for this. It's a formula that doesn't work.

단지 로렌츠 변환식을 변형해서 분모로 활용하였을 뿐 그 어떠한 이유도 찾을 수가 없다.

I just modified the Lorentz transformation and used it as the denominator, but I can't find any reason.

이 공식이 아인슈타인에 의해 수정한 공식이라고 알려졌는데 그것이 사실이라면, 이 공식은 기존 뉴턴역학의 가속도의 법칙에 관성을 적용하지 못한 그래프보다도 더 현실을 반영하지 못하는 완전히 잘못된 그래프를 그리게 된다.

It is known that this formula was modified by Einstein, and if that is true, this formula draws a completely wrong graph that does not reflect reality even more than the graph that fails to apply inertia to the acceleration law of Newtonian mechanics.

즉, 이 공식은 자연계의 법칙인 관성의 법칙에 위배된다.

In other words, this formula violates the law of inertia, which is a law of the natural world.

(c)의 그래프를 설명하기에 앞서 (d)의 공식과 그래프를 우선 살펴보자.

Before explaining the graph in (c), let's first look at the formula and graph in (d).

(d)의 그래프를 그린 공식은 아인슈타인의 특수상대성이론 중 질량 증가공식에 의하여 발생되는 그래프이다.

The formula that draws the graph in (d) is a graph generated by the mass increase formula in Einstein's special theory of relativity.

특수상대성이론의 관성질량 증가공식은 그래프에서 보다시피 광속에 이르러서 절대로 광속을 벗어날 수 없다는 것을 수치의 조합인 그래프로 증명하고 있다.

As you can see from the graph, the inertial mass increase formula of the special theory of relativity proves through a graph, which is a combination of numbers, that once you reach the speed of light, you can never escape the speed of light.

그러나, 앞서 가속도의 법칙에 로렌츠변환식을 첨가하여 수정한 (b)의 그래프에서 보았듯이 로렌츠변환식 자체가 관성의 법칙에 의한 질량 증가현상을 위반하고 있기 때문에 이 공식은 정확할 수 없다.

However, as seen in the graph in (b), which was modified by adding the Lorentz transformation equation to the law of acceleration, this formula cannot be correct because the Lorentz transformation equation itself violates the phenomenon of mass increase by the law of inertia.

몇 번을 말했듯이 질량을 가진 물체에 관성은 누적(적립)된다. 우리가 플라이 휠에 에너지를 저장했다가 사용할 수 있는 것도, 모두 관성이 누적되기 때문이다.

As I have said several times, inertia is accumulated in objects with mass. The reason we can store and use energy in a flywheel is because inertia accumulates.

이것을 운동에너지라고 주장하고 싶은 사람들이 분명 있을 것이다. 이 부분은 뒤에 자연스럽게 다루게 된다.

There will definitely be people who want to claim that this is kinetic energy. This part will be dealt with naturally later.

그러므로 (d)의 그래프도 관성의 법칙을 위반하고 있는 것이다.

Therefore, the graph in (d) also violates the law of inertia.

이제 마지막으로 (c)의 공식과 그래프를 살펴보자. (c)의 그래프와 공식은 아인슈타인이 특수상대성 이론에서 이야기하였던 전체 에너지와 관련된 그래프이다.

Now, finally, let's look at the formula and graph in (c). The graph and formula in (c) are graphs related to the total energy that Einstein talked about in the theory of special relativity.

즉, 전체 에너지란, $E = mc^2 + \frac{1}{2}mv^2$ 이다.

In other words, the total energy is $E = mc^2 + \frac{1}{2}mv^2$

이 공식을 있는 그대로 해석을 하자면, 핵융합이나 핵분열로 사용할 수 없는 어떠한 질량체라 하더라도, 그 질량이 가지는 에너지는 단지 인간이 사용하지 못할 뿐이지 에너지의 결정체인 것은 맞는다는 것이다.

If we interpret this formula as it is, even if it is a mass that cannot be used through nuclear fusion or fission, the energy of that mass cannot be used by humans, but it is true that it is a crystal of energy.

즉, 중력질량으로서 $mc^2$에 의하여 만들어진 물질의 운동으로 인하여 만들어지는 에너지가 있을 수 있으며 이 두 가지를 더하면 전체 에너지가 된다는 것이다.

In other words, there may be energy created by the movement of matter created by $mc^2$ as gravitational mass, and adding these two together becomes the total energy.

여기서 $mc^2$에 의해 만들어진 물질은 인간이 쓸 수 없는 에너지이자 질량이고, 질량이 운동함으로 인하여 만들어진 관성질량이 가지는 에너지는 인간이 쓸 수 있는 에너지이다.

Here, the material created by $mc^2$ is energy and mass that cannot be used by humans, and the energy possessed by the inertial mass created by the movement of the mass is energy that can be used by humans.

즉 $mc^2$에 의해 질량이 되어버린 에너지는 중력질량 역할만 하는 인간이 사용할 수 없는 에너지이고, 중력질량의 운동에너지 공식에 의해 발생하는 운동에너지는 인간이 사용할 수 있는 에너지이다.

In other words, the energy that becomes mass through

$mc^2$ is energy that cannot be used by humans, who only act as gravitational mass, and the kinetic energy generated by the kinetic energy formula for gravitational mass is energy that can be used by humans.

따라서 전체 에너지는 인간이 직접 사용할 수 없는 질량 역할만 하는 핵반응이 끝난 질량으로서의 에너지와 질량체의 운동으로 인하여 인간이 사용할 수 있는 관성질량에 의하여 생성되는 운동에너지를 더한 값이다.

Therefore, the total energy is the sum of the energy as the mass after the nuclear reaction, which only acts as a mass that cannot be used directly by humans, and the kinetic energy generated by the inertial mass that can be used by humans due to the movement of the mass body.

그러나 이 공식에 의한 전체 에너지의 그래프도 관성에 의한 질량 증가현상을 적용한 가속도 그래프(a')와 다른 모습으로 나타나고 있다.

However, the graph of total energy based on this formula also appears different from the acceleration graph (a') that applies the mass increase phenomenon due to inertia.

우선 이 공식은 $mc^2$에 운동에너지인 $\frac{1}{2}mv^2$을 한번 더하기를 했다는 점에 대하여 살펴봐야 한다.

First of all, we need to look at the fact that this formula

adds $\frac{1}{2}mv^2$, the kinetic energy, to $mc^2$.

아인슈타인은 '질량은 에너지다.'라는 등가관계를 성립시킨 인물이다.

Einstein is the person who established the equivalence relationship: 'Mass is energy.'

그런 면에서 중력질량에 운동에너지를 더하면 전체 에너지가 된다는 발상은 참신하다.

In that respect, the idea that adding kinetic energy to gravitational mass creates total energy is novel.

그러나, 운동에너지 공식이 질량이 일정할 때라는 전제를 조건으로 만들어진 것이라는 사실을 간과했다.

However, the fact that the kinetic energy formula was created based on the premise that the mass is constant was overlooked.

$W = F \times S = m \times a \times s = m \times ((V - VO)/t) \times ((V + VO)/2 \times t) = m \times (V\char`^2 - VO\char`^2)/2 = m \times V\char`^2/2$

상기 식에서 보듯이 $\frac{1}{2}mv^2$ 이 $F = ma$에서 유도된 것을 알 수 있게 해준다.

As seen in the above equation, it allows us to see that $\frac{1}{2}mv^2$ is derived from $F = ma$.

결국 질량이 일정할 때를 전제조건으로 일을 구하는 공식에 의하여 운동에너지 공식이 성립되었음을 알 수 있다.

In the end, it can be seen that the kinetic energy formula was established through the formula for calculating work under the precondition that the mass is constant.

그런데 앞서 F=ma 공식에 있어서 질량이 일정할 수 없음을 알았다. (누적되는 관성에 의해서 질량이 일정할 수가 없다.)

However, we learned earlier that mass cannot be constant in the formula F=ma. (Mass cannot be constant due to cumulative inertia.)

그러므로 이 운동에너지 공식도 오류를 가지고 있고, 그에 따라 아인슈타인의 전체 에너지 공식도 오류를 가지고 있다.

Therefore, this kinetic energy formula also has an error, and accordingly, Einstein's total energy formula also has an error.

그런데 아인슈타인은 왜? 여기에 중력질량에 해당하는 값을 더해서 전체 에너지라는 공식을 만들었을까?

But did Einstein create a formula for total energy by adding the value corresponding to gravitational mass?

그것은, 기존의 운동에너지 공식이나 그 값이 에너지를 충분히 설명하지 못하고 있다는 것을 아인슈타인이 알고 있었다는 것을 방증하는 것은 아닐까?

Doesn't this prove that Einstein knew that the existing kinetic energy formula or its value did not sufficiently explain energy?

아인슈타인의 시도는 매우 좋았다. 그러나, 그 전제인 질량이 일정할 수 없다는 것을 결국 이해하지 못하였다.

Einstein's attempt was very good. However, in the end, it was not understood that mass cannot be constant.

그는 질량이 일정할 때라는 전제로 설명된 운동에너지 공식을 그냥 가져다 쓰는 오류를 범하고 말았다.

He made the mistake of simply using the kinetic energy formula, which was explained with the assumption that the mass was constant.

이 글을 시작할 때, (a), (b), (c), (d) 네 개의 공식과 그래프는 서로 연관성이 있는 것들이라고 밝힌 바 있다.

At the beginning of this article, I stated that the four formulas and graphs (a), (b), (c), and (d) are related to each other.

그렇다. 이것들은 모두 관성과 그 관성으로 인한 관성질량 증가현상을 표현하기 위한 아인슈타인과 뉴턴의 노력이라고 말할 수 있다.

Yes. These can all be said to be Einstein and Newton's efforts to express inertia and the phenomenon of inertial mass increase due to that inertia.

그들은 관성을 표현하기 위해 부단히 노력하였지만, 결과는 좋지 않았다.

They tried hard to express inertia, but the results were not good.

아인슈타인의 질량 증가공식 및 뉴턴역학의 가속도의 법칙을 수정한 수식과 전체 에너지 공식 모두가 관성의 법칙으로 인한 관성질량 증가현상을 표현하지 못하고 있다.

Einstein's mass increase formula, the formula that modified Newton's law of acceleration in mechanics, and the total energy formula all fail to express the phenomenon of inertial mass increase due to the law of inertia.

물론 아인슈타인의 관성질량 증가공식에 의한 그래프는 처음부터 관성의 법칙에 의한 관성질량 증가현상과 다른 그래프의 방향을 보여주고 있다.

Of course, the graph based on Einstein's inertial mass increase formula shows a different direction from the beginning than the inertial mass increase phenomenon based on the law of inertia.

그리고, 뉴턴은 가속도의 법칙에서 관성질량 증가현상을 적용하지 못하는 오류를 범하였다. (관성질량 증가현상은 관성의 법칙에 의하여 발생한다.)

And, Newton made the error of failing to apply the phenomenon of inertial mass increase in the law of acceleration. (The phenomenon of inertial mass increase occurs due to the law of inertia.)

시작 질량이 0인 물체는 관성질량 증가가 일어나지 않고, 시작 질량이 0보다 큰 물체의 가속도는 힘에 비례하지 않고, 증가된 관성질량만큼 가속도의 증가량이 줄어든다.

For objects with a starting mass of 0, the inertial mass does not increase, and the acceleration of an object with a starting mass greater than 0 is not proportional to the force, and the increase in acceleration is reduced by the increased inertial mass.

힘에 의해 관성질량이 생기기 때문이다.

This is because inertial mass is created by force.

그러므로 4개의 그래프와 수식은, 모두 오류를 가지고 있다.

Therefore, all four graphs and formulas contain errors.

다만, 수식으로 표현되지 않은 운동법칙인 뉴턴의 제1법칙의 관념만이 관성을 정확하게 표현하고 있을 뿐이다.

However, only the concept of Newton's first law, which is a law of motion that is not expressed in formulas, accurately expresses inertia.

"외부로부터 힘이 작용하지 않으면 물체의 운동 상태는 변하지 않는다."

"The state of motion of an object does not change unless an external force acts on it."

어쩌면 인간은 '가속도의 법칙에 관성이 적용되지 않은 그래프는 큰 문제가 아닐 것'이라고 생각했을 수도 있다.

Maybe humans thought, 'A graph where inertia is not applied to the law of acceleration would not be a big problem.'

그러나 이러한 사소한 오류가 관성의 법칙과 엔트로피 법칙이 충돌을 일으킨다는 사실을 감추게 될 수도 있는 것이다.

However, these minor errors may hide the fact that the law of inertia and the law of entropy are in conflict.

관성질량이 증가하는 이유는 관성의 법칙이 적용되는 관성계라는 뜻이고, 관성질량은 질량의 증가가 멈출 때까지 가속도와 같이 증가하며 광속에 가까울수록 질량의 증가속도와 가속도의 증가속도는 같이 느려지다가 결국 평형을 이루는 것이 옳을 것이다.

The reason the inertial mass increases is because it means that it is an inertial system where the law of inertia applies. The inertial mass increases along with the acceleration until the increase in mass stops, and as it approaches the speed of light, the rate of increase in mass and the rate of increase in acceleration slow down. In the end, it would be right to achieve equilibrium.

그러나 지금까지 물리학은 광속에 가까울수록 관성질량이 무한대가 된다고 인류를 학습시키고 있다.

However, until now, physics has taught humanity that the closer you get to the speed of light, the inertial mass becomes infinite.

속도의 증가와 관성질량의 증가가 평형 상태가 될 때까지만 질량 증가가 일어나며, 광속에 가까워지기 전에 이미 (누적된) 관성질량 때문에 무거워진 질량체는 광속에 가까울수록 관성질량을 증가시키기 어렵다. 따라서 관성질량 증가량이 점점 줄어드는 것을 설명하지 못하고 있다. (같은 힘으로 큰 질량체와 작은 질량체를 움직이는 실험을 통해 작은 질량체를 움직이는 것이 더 쉽다는 것은 누구나 아는 일이다.)

Mass increase occurs only until the increase in speed and the increase in inertial mass reach an equilibrium state, and for a mass that has already become heavier due to (accumulated) inertial mass before approaching the speed of light, it is difficult to increase the inertial mass the closer it approaches the speed of light. Therefore, it cannot explain the gradual decrease in the increase in inertial mass. (Everyone knows through experiments of moving a large mass and a small mass with the same force that it is easier to move a small mass.)

이것이, 관성의 법칙에 의한 관성질량 증가방식과 방식이 다른 로렌츠인자를 받아들인 결과다.

This is the result of accepting the Lorentz factor, which is different from the method of inertial mass increase according to the law of inertia.

# 뉴턴의 가속도의 법칙의 흠결

## Flaws in Newton's law of acceleration

지금까지 우리가 배워온 가속도의 법칙 그래프는 시작질량이 0일 때에만 성립한다. 그러나 시작질량이 0인 물체는 관성의 법칙이 적용되지 않는다.

The acceleration law graph we have learned so far is valid only when the starting mass is 0. However, the law of inertia does not apply to objects with a starting mass of 0.

시작질량이 0이면 질량 증가값도 0이기 때문이다.

This is because if the starting mass is 0, the mass increase value is also 0.

따라서, 시작질량이 0이 아니면서 0보다 큰 질량값을 가지고 있을 때, 가속도의 법칙 그래프는 일정한 기울기를 가져야 한다. (가해지는 힘에 의하여 관성질량이 증가하기 때문에 속도의 증가량이 그만큼 줄어든다.)

Therefore, when the starting mass is not 0 and has a mass value greater than 0, the acceleration law graph must have a constant slope. (Because the inertial mass increases due to the applied force, the increase in speed decreases accordingly.)

그러므로, 시작질량값이 0이면 질량 증가가 일어나지 않고, 시작질량값이 0보다 크면 관성의 법칙 때문에 질량 증가가 일어나고 가속도의 법칙 그래프는 늘어나는 관성질량만큼 기울기를 가지게 된다.

Therefore, if the starting mass value is 0, mass increase does not occur, and if the starting mass value is greater than 0, mass increase occurs due to the law of inertia, and the acceleration law graph has a slope corresponding to the increasing inertial mass.

그러므로, 현재 우리가 알고 있는 가속도의 법칙 그래프는 시작질량이 0인 경우에만 성립하는 그래프이고, 시작질량이 0보다 큰 일정한 질량을 가진 경우에는 맞지 않는 그래프이다.

Therefore, the acceleration law graph that we currently know is a graph that holds true only when the starting mass is 0, and it is a graph that does not fit when the starting mass has a constant mass greater than 0.

시작질량이 0보다 큰 일정한 질량을 가진 입자는 굳이 상대성 이론의 질량 증가이론이 아니더라도, 늘어나는 관성질량의 양 때문에 광속을 벗어날 수 없다. (광속에 가까울수록 누적된 관성질량으로 인하여 무한대의 힘을 가해도 속도를 증가시키기 어려워지고 더 이상의 관성질량을 증가시키기도 어려워질 것이기 때문이다.)

Particles with a constant mass whose starting mass is greater than 0 cannot escape the speed of light due to the increasing amount of inertial mass, even if it is not a mass increase theory of relativity theory. (This is because the closer it is to the speed of light, the more difficult it will be to increase the speed due to the accumulated inertial mass even if infinite force is applied, and it will be difficult to increase the inertial mass any further.)

이것이 가속도의 법칙이 가지고 있는 가정의 오류 또는 흠결이다.

This is an error or flaw in the assumption of the law of acceleration.

0보다 큰 시작질량을 가진 입자에 힘(F)을 가하면, 관성의 법칙이 작용하여 관성질량이 증가하기 때문에 시작질량보다 질량값이 커진다. 그렇기 때문에 질량이 일정할 수 있는 방법이 없다.

When a force (F) is applied to a particle with a starting mass greater than 0, the law of inertia acts and the inertial mass increases, so the mass value becomes larger than the starting mass. Therefore, there is no way for mass to be constant.

# 아인슈타인의 질량 증가이론과 질량 증가공식의 흠결
# Flaws in Einstein's mass increase theory and mass increase formula

'빛은 질량이 없다'는 주장은 빛이 빛으로 불리우는 동안은 입자이자 파동이라 말할 수 있지만 질량체다.

The claim that 'light has no mass' means that while light is called light, it can be said to be a particle or a wave, but it is a mass.

다만, 여기서 광자의 질량까지 논할 필요는 없다. 단지, 질량이 0보다 큰 입자의 질량 증가와 질량 증가공식을 이야기하고자 함이다.

However, there is no need to discuss the mass of photons here. I just want to talk about the mass increase of particles with a mass greater than 0 and the mass increase formula.

특수상대성이론 중 시간지연효과, 길이수축효과, 질량증가효과 중에, 시간지연이나 길이수축은 관성의 법칙과 인과관계가 없다.

Among the time delay effect, length contraction effect, and

mass increase effect in special relativity, time delay or length contraction have no causal relationship with the law of inertia.

시간지연효과나 길이수축효과를 설명하는 공식이 관성의 법칙을 어기던 말던 관심이 없다.

I am not interested in whether the formulas that explain the time delay effect or the length contraction effect violate the law of inertia.

그것들은 질량을 가지지 않는 물리현상이기 때문에 관성의 법칙의 적용을 받지 않는다.

Because they are physical phenomena that do not have mass, they are not subject to the law of inertia.

그러나, 질량증가이론이나 그 효과는 질량을 가진 물체에 작용하는 관성의 법칙과도 상호 보완적이어야 한다.

However, the theory of mass increase and its effects must be complementary to the law of inertia that acts on objects with mass.

뉴턴역학의 가속도의 법칙의 가정의 오류를 찾아내지 못한 아인슈타인은 F=ma 공식에 대하여, 힘을 가하면, 광속을 벗어날 수 있다고 생각하는 공식으로 인식되고, 이 공식은 오류가 있다고 생각하게 된다.

Einstein, who could not find the error in the assumption of the law of acceleration in Newtonian mechanics, recognized the F=ma formula as a formula that believed that applying force could escape the speed of light, and came to think that this formula was in error.

이런 생각은 아인슈타인으로 하여금 뉴턴의 가속도의 법칙 공식을 수정하게 만든다.

This idea led Einstein to modify Newton's law of acceleration formula

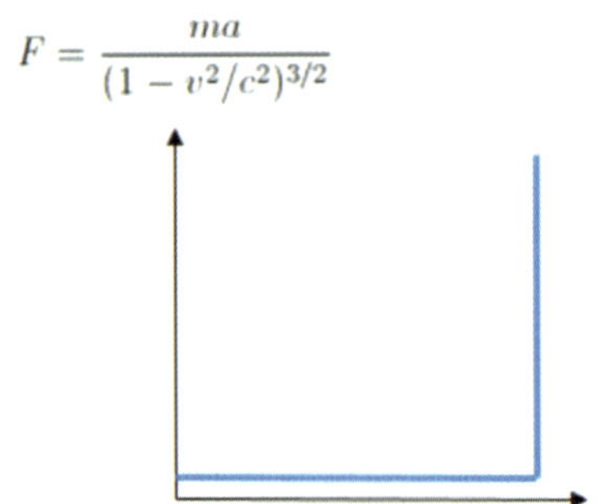


(b) 아인슈타인 수정 가속도의 법칙

(b) Einstein's modified law of acceleration

이 수정공식은 단지 로렌츠변환식이 적절하게 변환되어 분모로 작용한다는 것이었다.

This correction formula simply means that the Lorentz transformation equation is properly converted and acts as the denominator.

아인슈타인의 질량증가이론은 결국 로렌츠변환식을 차용하는 오류를 범하게 되었다. 이로 인하여 질량증가 그래프가 잘못 작성되었다. 이런 이유로 질량 증가이론은 맞을지라도 질량 증가공식은 관성의 법칙을 무시한 잘못 작성된 공식이 되었다.

Einstein's theory of mass increase ultimately made the error of borrowing the Lorentz transformation equation. Because of this, the mass increase graph was drawn incorrectly. For this reason, although the mass increase theory is correct, the mass increase formula is a poorly written formula that ignores the law of inertia.

질량증가공식이 잘못된 것은 철저하게 관성의 법칙을 적용을 받는 질량이 있는 물질에 대한 질량 증가에 있어서, 관성의 법칙을 철저하게 무시하는 로렌츠변환식을 차용해왔기 때문이다.

The reason the mass increase formula is wrong is because the Lorentz transformation formula, which completely ignores the law of inertia, has been adopted to increase mass for substances with mass that are strictly governed by the law of inertia.

# 기초물리학의 흠결 결여
## Resolving the flaws in basic physics

이 두 명의 천재 과학자들의 가정의 오류와, 공식의 오류로 인간은 에너지를 창조하는 존재로 진화하는 문 앞에서 좌절을 경험하고 있는 것으로 생각한다.

I believe that due to the errors in the assumptions and formulas of these two genius scientists, humans are experiencing frustration on the doorstep of evolving into beings who create energy.

관성의 법칙은 외력이 없을 경우 물체는 항상 등속직선운동 상태, 즉, 일직선을 따라 같은 속력으로 움직이는 상태를 유지한다는 법칙이다.

The law of inertia is the law that states that in the absence of external force, an object always maintains a state of constant linear motion, that is, moving at the same speed along a straight line.

이때, 외력이 운동방향과 같은 진행방향으로 가해지면, 그 외력의 일부는 물체의 질량도 증가시키고 가속도도 증가시킨다.

At this time, when an external force is applied in the same direction as the direction of movement, part of the external force increases the mass of the object and also increases its acceleration.

그래서 같은 진행방향으로 그리고 일정한 단위로 외력이 가해지면 질량 증가가 일어난 만큼 가속도의 증가량은 조금씩 줄어든다. 이 현상은 그래프의 기울기로 표현된다.

So, if an external force is applied in the same direction and in a certain unit, the increase in acceleration gradually decreases as the mass increases. This phenomenon is expressed as the slope of the graph.

# 운동에너지
Kinetic energy

우리는 다양한 운동에너지 공식을 지금까지 사용하여 왔다.

We have used various kinetic energy formulas so far.

예를 들어 질점에 관한 운동에너지는 입자의 질량이 일정하다고 하고 그 입자에 행해진 총 일은 뉴턴의 두 번째 법칙에 의해 $T = \frac{m|v|^2}{2}$ 으로 정의된다.

For example, the kinetic energy of a particle is assumed to have a constant mass of the particle, and the total work done on the particle is defined as $T = \frac{m|v|^2}{2}$ by Newton's second law.

그러므로, 직선운동에 있어서의 운동에너지는 $E = \frac{1}{2}mv^2$ 으로 정의된다.

Therefore, the kinetic energy in linear motion is defined as $E = \frac{1}{2}mv^2$.

또한 회전 운동의 운동에너지는 관성모멘트와 가속도의 제곱에 비례하며 $E = \frac{1}{2} Iw^2$ 로 정의된다.

Additionally, the kinetic energy of rotational motion is proportional to the square of the moment of inertia and angular velocity and is defined as $E = \frac{1}{2} Iw^2$.

이때 I는 관성 모멘트이고 $\omega$는 가속도이다. 주어진 축을 중심으로 회전하는 점질량에 대한 스칼라 관성모멘트는 $I = mr^2$로 정의된다.

At this time, I is the moment of inertia and $\omega$ is the angular velocity. The scalar moment of inertia for a point mass rotating about a given axis is defined as $I = mr^2$.

즉, 이러한 운동에너지를 정의함에 있어서, 모든 입자의 질량이 일정하다고 전제하고, 뉴턴의 두 번째 법칙인 가속도의 법칙에 근거해서 운동에너지 공식이 정해져 있음을 볼 수 있다.

In other words, in defining this kinetic energy, it is assumed that the quantity of all particles is constant, and it can be seen that the kinetic energy formula is established based on Newton's second law, the law of acceleration.

그러나, 앞서 밝혔듯이, 질량이 일정할 수 있는 방법은, 정지해 있지 아니한 상태에서는 질량이 0인 경우 이외에는 불가능하다. 또 정지해 있는 경우는 질량이 가해진 힘으로 움직일 수 없을 만큼 큰 경우가 아니면 질량은 일정할 수 없는 것이다.

However, as stated earlier, there is no way for the mass to

be constant except when the mass is 0 in a non-stationary state. Also, when it is at rest, the mass cannot be constant unless it is so large that it cannot be moved by the applied force.

그러므로 실제 운동에너지는 기존 운동에너지 공식에서 나온 값보다는 항상 크다.

Therefore, the actual kinetic energy is always greater than the value derived from the existing kinetic energy formula.

# 엔트로피 법칙과 관성의 법칙
# Law of Entropy and Law of Inertia

열역학 제2법칙이라 불리는 엔트로피 법칙에 대해서 알아본다.

Learn about the law of entropy, also known as the second law of thermodynamics.

에너지 측면에서 엔트로피의 증가는 인간이 사용할 수 있는 에너지가 줄어드는 것을 뜻한다고 할 수 있다.

In terms of energy, an increase in entropy means a decrease in the energy available to humans.

롤러코스터에서 움직이는 중력에 의한 퍼텐셜 에너지가 운동에너지로 변환되는 것은 계가 바뀌는 것이다. 다만, 그 위치까지 그 질량을 이동시키는 에너지가 더 많이 소요되었을 뿐이다.

When the potential energy due to gravity moving on a roller coaster is converted into kinetic energy, the system changes. However, it only took more energy to move the mass to that location.

화약의 화학에너지가 총알의 운동에너지로 변환되는 것 또한 계가 바뀌는 것이다.

The conversion of the chemical energy of gunpowder into the kinetic energy of a bullet also changes the system.

이 또한, 그 계가 바뀔 때마다 인간의 관점에서 사용할 수 있는 에너지는 줄어들고 사용할 수 없는 에너지가 늘어나는 과정을 거치면서, 우리는 에너지 보존의 법칙이 맞는다고 인식하게 된 것이다.

In addition, as the system changes, from a human perspective, usable energy decreases and unusable energy increases, we have come to recognize that the law of conservation of energy is true.

이것을 엔트로피 법칙이라 하여 열역학 제2법칙으로 설명하고 있다.

This is called the law of entropy and is explained by the second law of thermodynamics.

이 엔트로피 법칙은 고전역학의 뉴턴의 운동법칙을 바탕으로 만들어졌다고 한다.

It is said that this entropy law was created based on Newton's laws of motion in classical mechanics.

그래서 처음 과학자들은 기존의 고전역학처럼 시간의 방향성을 설명해주지 못한다고 판단했다.

So, at first, scientists decided that it could not explain the direction of time like existing classical mechanics.

그러나 과학자들은 엔트로피 법칙의 역의 경우가 물리법칙에 위배되지 않는다는 말을 부정하기 위해 과거로 거슬러 올라가 '빅뱅' 이론에서부터 출발하여 엔트로피 법칙이 옳음을 주장한다.

However, in order to deny that the inverse case of the law of entropy does not violate the laws of physics, scientists go back to the past, starting from the 'Big Bang' theory, and claim that the law of entropy is correct.

즉, 태초의 우주로 도달하게 되면, 이 엔트로피 법칙의 역은 성립되지 않는다는 사실을 너무나 명쾌하게 알 수 있으며, 엔트로피 법칙의 역은 '우주의 법칙에 순응하지 않는 역행'이며 결코 일어날 수 없는 것이라 결론지었다.

In other words, when we reach the primordial universe, we can clearly see that the inverse of the entropy law does not hold, and we conclude that the inverse of the entropy law is a 'reverse that does not comply with the laws of the universe' and can never happen. Built.

인간이 사용할 수 있는 에너지가 감소하면서 발생한 엔트로피의 증가가 압도적으로 많다는 것이고, 저-엔트로피를 만드는 순간에도 물질계는 고-엔트로피로 향했다는 이야기다.

The increase in entropy that occurred as the energy usable by humans decreased was overwhelming, and even at the moment of creating low entropy, the material world was heading toward high entropy.

엔트로피 법칙의 역의 경우가 만에 하나 맞는다고 가정하면, 우주는 점점 수축해가고 있어야 한다고도 한다.

Assuming that the inverse of the law of entropy is true, it is also said that the universe should be gradually shrinking.

엔트로피가 고-엔트로피로 되는 과정, 즉 물질계가 분산되는 과정에서는 사용 불가능한 에너지가 필연적으로 발생하게 되며, 만일 엔트로피 법칙이 역전될 수도 있다면 고-엔트로피가 되는 과정에서 발생하는 사용 불가능한 에너지를 역전과정에서 100% 재순환을 시켜 다시 사용 가능한 에너지로 바꿔야 한다는 소리다.

In the process of entropy becoming high-entropy, that is, the process of dispersion of the material world, unusable energy is inevitably generated, and if the law of entropy can be reversed, the unusable energy generated in the process of becoming high-entropy must be used in the reversal process. This means that we need to 100% recycle it and turn it into usable energy again.

그러나, 이러한 에너지 변화과정은 인간의 인위적인 힘으로는 절대 불가능하고, 게다가 이 과정은 자연적으로 이루어져야 하는데 자연 역시 이미 사용이 불가능해진 에너지를 사용이 가능한 에너지로 변환시키는 힘은 존재하지 않는다고 그동안 과학자들은 설명했다.

However, this process of energy change is absolutely impossible with human artificial power, and in addition, this process must be done naturally, but scientists have explained that nature also does not have the power to convert already unusable energy into usable energy. did.

에팅턴은 "엔트로피 법칙은 자연법칙 가운데 최고의 법칙이다."라고 말을 하였고, 아인슈타인은 "엔트로피는 붕괴되지 않을 유일한 이론이다."라는 말을 남겼다고 한다.

Ettington said, "The law of entropy is the best of the laws of nature." Einstein said, "Entropy is the only theory that will not collapse." It is said that he left a comment.

이제 엔트로피 법칙을 좀 더 세밀하게 정의해보면, 엔트로피 법칙이란 물질과 에너지는 하나의 방향으로만, 즉 사용이 가능한 것에서 불가능한 것으로, 또는 이용이 가능한 것에서 불가능한 것으로 또는 질서 있는 것에서 무질서한 것으로 변화하는 것으로 정의된다.

Now, if we define the law of entropy in more detail, the law of entropy is defined as the fact that matter and energy change in only one direction, that is, from usable to unusable, or from usable to unusable, or from ordered to disordered.

이 법칙에 의하면 에너지 변환이 이루어질 때마다 인간이 사용할 수 있는 에너지는 줄어들고, 사용할 수 없는 에너지는 늘어나는 것을 엔트로피 법칙이라는 이름으로 정의된다.

According to this law, whenever energy conversion occurs, the energy that humans can use decreases, and the energy that cannot be used increases, which is defined as the law of entropy.

다시 정리해보면,
인간의 관점에서 에너지는
(1) 사용 가능한 것에서 불가능한 것으로 변화하고,
(2) 이용 가능한 것에서 불가능한 것으로 변화하며,
(3) 질서가 있는 것에서 무질서한 것으로 변화하는
것으로 인식해온 것이고 그것을 엔트로피의 법칙이라고 부른 것이다.

To summarize again,
From a human perspective, energy has been recognized as
(1) changing from usable to unusable,
(2) changing from usable to unavailable, and
(3) changing from ordered to disordered.
This is called the law of entropy.

이 법칙은 대단히 인간 중심의 법칙으로 여겨진다.
This law is considered a very human-centered law.

인간에게 있어서 에너지 소비의 결과물인 이산화탄소($CO_2$)는 인간의 입장에서 보면, 산소를 소비해서 나온 2차 물질이고, 더 이상 사용할 일이 없는 것으로 여겨진다.

From the human perspective, carbon dioxide ($CO_2$), which is a result of energy consumption for humans, is a secondary substance produced by consuming oxygen and is considered to have no further use.

그러나, 녹색식물의 입장에서는 광합성의 연료이고, 이 광합성을 통해 탄소를 사용하며, 인간이 사용할 수 있는 산소를 배출한다.

However, from the perspective of green plants, it is a fuel for photosynthesis, and through this photosynthesis, it uses carbon and releases oxygen that humans can use.

인간에게는 저-엔트로피인 산소가, 광합성 식물에게는 사용하고 쓸모없어진 고-엔트로피 물질인 것이다.

Oxygen, which is low-entropy for humans, is a high-entropy substance that has been used and become useless for photosynthetic plants.

다시 말해 순환과정이 그려진 것이다.

In other words, a circular process is depicted.

"개개인의 물질적 부의 축적이 사회를 발전시키는 일이다."라고 주장한 로크는 "자연계에 전적으로 맡겨진 땅은 쓰레기다."라는 말을 남겼다.

Locke argued that "the accumulation of individual material wealth is what develops society." He said, "Land left entirely to the natural world is trash."

자연계에 잡초가 무성하니 쓰레기라고 할 만할지 모르겠다.

Since there are so many weeds in the natural world, I don't know if it can be called trash.

그러나, 자연계에 맡겨진 땅은 잡초가 무성하게 생겨나고, 광합성을 해서, 이산화탄소를 원료로 산소와 탄소로 분리하고, 다시 잡초가 썩어 해당 탄소가 땅을 튼튼하게 하고, 다시 곡식을 통해 우리가 섭취하게 한다는 순환의 원리를 이해하지 못한 듯하다.

However, in the land left to the natural world, weeds grow abundantly, photosynthesis occurs, and carbon dioxide is separated into oxygen and carbon as raw materials. Then, the weeds rot and the carbon strengthens the land, which is then consumed by us through grain. It seems like you don't understand the principle of circulation.

산림이나 들판의 풀과 나무를 태워버리고 그 자리에 곡식 등을 재배하는 화전농에 있어서 자연의 역할을 완벽하게 무시한 발언이라 할 수 있다.

This can be said to be a statement that completely ignores the role of nature in slash-and-burn farming, which involves burning down grass and trees in forests or fields and growing grains in their place.

인간의 관점에서 보면 쓸모없는 고-엔트로피 물질일지 모르지만, 자연의 관점에서 보면 그 고-엔트로피 물질이 저-엔트로피의 원료로 치부될 수 있는 것이다.

From a human perspective, it may be a useless high-entropy material, but from a natural perspective, the high-entropy material can be dismissed as a low-entropy raw material.

그렇다면, 엔트로피 법칙의 현실적인 역은 어떻게 설명될까?

So, how can the realistic inverse of the entropy law be explained?

인간의 관점에서 보는 엔트로피의 역은,

The inverse of entropy from a human perspective is:

(1) 이용이 불가능한 것에서 가능한 것으로 변화되거나,

(1) Changes from unusable to possible,

(2) 사용이 불가능한 것으로 인식되어 오던 것에서 가능한 것으로 새롭게 인식되거나,

(2) What was once recognized as impossible to use is newly recognized as possible, or

(3) 무질서한 것에서 질서 있는 것으로 변화하게 되는 것을 말한다고 할 수 있다.

(3) It can be said to refer to a change from disorder to order.

이 내용을 잘 살펴보면 자연계의 법칙 중 관성의 법칙과 일치한다.

If you look at this carefully, it is consistent with the law of inertia among the laws of the natural world.

뉴턴이 제시한 관념인 관성의 법칙의 문장 이후에 이보다 더 관성을 정확하게 표현하기는 어렵다.

It is difficult to express inertia more accurately after the sentence of the law of inertia, which is an idea presented by Newton.

$E = mc^2$의 등식에 따라 핵융합 또는 핵분열 반응이 모두 끝난 어떤 핵융합이나 핵분열로는 더 이상 사용할 수 없는 질량체가 있다고 가정하자.

Let us assume that there is a mass that can no longer be used in any nuclear fusion or fission reactor where the nuclear fusion or fission reaction has completed according to the equation $E = mc^2$.

아인슈타인의 전체 에너지의 개념은 이 질량체를 움직였을 때, 질량체 자체가 가지고 있는 질량과 그 운동에너지의 합이다.

Einstein's concept of total energy is the sum of the mass of the mass itself and its kinetic energy when this mass is moved.

이 물질은 아인슈타인의 전체 에너지 이론에 의해 중심축이 있는 회전운동을 통해 관성(관성모멘트)을 얻어낼 수 있다. 질량이 있기 때문이다.

This material can obtain inertia (moment of inertia) through rotational movement around a central axis according to Einstein's total energy theory. Because it has mass.

(단지 아인슈타인의 질량 증가공식이나 전체 에너지 공식이 틀렸다고 해서 그 관념까지 틀린 것은 아니다. 수학적 공식의 오류와 과정의 오류일 뿐이다.)

(Just because Einstein's mass increase formula or total energy formula is wrong does not mean that the concept is wrong. It is just an error in the mathematical formula and an error in the process.)

이것은 또 사용이 불가능하다고 인식되어 오던 것에서 사용이 가능한 것으로 새롭게 인식되었다고 볼 수도 있다.

This can also be seen as a new recognition of something that can be used from something that was previously perceived as impossible to use.

뿐만 아니라 관성질량이 누적되는 관성질점은 nF의 힘이 주어졌을 때, $2^n-1$의 개수만큼 규칙적으로 증가한다.

In addition, the inertial point where the inertial mass is accumulated regularly increases by the number of $2^n-1$ when a force of nF is given.

자, 다시 작성하여 보면 관성의 법칙에 기인한 관성질량의 증

가는,

Now, if we rewrite it, the increase in inertial mass due to the law of inertia is:

(1) 모든 핵반응이 종료된 정지된 질량이 있는 물질을 이용하여 사용이 가능한 관성질량을 얻어낼 수 있으며,

(1) A usable inertial mass can be obtained by using a substance with a stationary mass in which all nuclear reactions have been completed.

(2) 관성질량은 가해진 힘(F)에 비례하는 것이 아니라, 가해진 힘의 방향을 따르나, 질량에 비례하여 사용 가능한 에너지로 바뀔 수 있으며,

(2) Inertial mass is not proportional to the applied force (F), but follows the direction of the applied force, but can be converted into usable energy in proportion to the mass.

(3) nF의 힘이 주어졌을 때, $2^n-1$개에 해당하는 가상질점과 가상질점이 가지는 관성력(관성질량)을 질서 있고 규칙적으로 만들어낸다.

(3) When a force of nF is given, $2^n-1$ virtual matter points and the inertial force (inertial mass) of the virtual matter points are created in an orderly and regular manner.

그러므로 관성의 법칙과 엔트로피 법칙은 서로 상반된 결과를

주장하는 이론이 된다.

Therefore, the law of inertia and the law of entropy are theories that claim conflicting results.

에너지가 변화되는 것은 결국 어떤 계에서 다른 계로의 변화되는 과정에서 이루어진다.

Energy changes ultimately occur in the process of changing from one system to another.

즉, 다양한 에너지가, 서로 변환되는 과정은 계가 바뀌는 과정이고 이 과정에서 우리는 에너지가 소비되는 것을 지금까지 목격하여 왔다.

In other words, the process of converting various energies into each other is a process of changing the system, and in this process, we have witnessed energy being consumed.

그러나 관성은 힘이 주어지면 주어진 힘인 에너지를 소비하는 과정에서 힘에 비례하는 관성을 증가시키는 것이 아니라, 질량에 비례하여 관성을 증가시킨다는 특이점을 가지고 있다.

However, inertia has the peculiarity that when a force is given, in the process of consuming energy, which is the given force, the inertia does not increase in proportion to the force, but inertia increases in proportion to the mass.

이 말은 초기 중력질량에 누적되는 관성질량을 더한 값은 초기 중력질량보다는 항상 더 많은 양이라는 뜻이 된다. (그러므로 질량이 늘어난 만큼 질량에너지 등가의 원리에 따라 에너지가 늘어나는 것이 자연계의 법칙에 부합하는 현상인 것이다.)

This means that the initial gravitational mass plus the accumulated inertial mass is always greater than the initial gravitational mass. (Therefore, as mass increases, energy increases according to the principle of mass-energy equivalence, a phenomenon that is in accordance with the laws of the natural world.)

이제 관성질량이 커졌으므로, 전체 에너지는 초기 에너지값보다 더 큰 에너지값을 가진다.

Now that the inertial mass has increased, the total energy has an energy value greater than the initial energy value.

우리는 늘어난 관성질량만큼의 에너지를 꺼내어 쓸 수 있는 방법을 고안하면 된다. (질량에너지 등가원리에 의해서 늘어난 관성질량은 모두 에너지이기 때문이다.)

We just need to devise a way to extract and use energy equal to the increased inertial mass. (This is because all inertial mass increased according to the mass-energy equivalence principle is energy.)

초기 질량이 가지고 있던 $mc^2$에 해당하는 관성의 양과 힘(nF)을 가해서 얻은 $2^n$-1개의 가상질점이 가지고 있는 관성질량을 더한 값은, 어떤 경우에도 초기질량이 가지고 있는 관성의 값보다 크다.

The value of adding the amount of inertia corresponding to $mc^2$ possessed by the initial mass and the inertial mass possessed by $2^n-1$ virtual masses obtained by applying force (nF) is greater than the value of inertia possessed by the initial mass in any case.

조금 더 관념을 확장해보자. 일반적으로 질량 증가에 대하여 부정할 사람은 없다. 다만, 필자는 아인슈타인이 세운 질량 증가 공식에 문제가 있었을 뿐이라고 생각한다.

Let's expand the concept a little further. In general, no one denies mass increase. However, I think there was just a problem with Einstein's mass increase formula.

특수상대성 이론의 시간팽창이나, 길이수축과 같이 질량과 관계없는 물리현상에 로렌츠변환을 적용하는 것이 옳은 것인지 여부와 관계없이, 로렌츠변환식은 관성의 법칙에 위배되는 방정식이기 때문에 질량 증가공식에 부적합한 방정식이었다.

Regardless of whether it is correct to apply the Lorentz transformation to physical phenomena unrelated to mass, such as time expansion or length contraction in the theory of special relativity, the Lorentz transformation equation is an equation that is unsuitable for the mass increase formula because it is an equation that violates the law of inertia. It was.

관성력은 관성을 힘으로 표현한 것이고, 관성질량은 관성을 질량으로 표현한 것이다.

Inertial force is the expression of inertia as force, and inertial mass is the expression of inertia as mass.

그러므로, 앞서 우리가 밝힌 바와 같이 질량을 가진 물체에 nF의 힘이 가해졌을 때, 가상질점은 $2^n$-1개만큼 발생되어 완벽한 질량 역할을 한다.

Therefore, as we revealed earlier, when a force of nF is applied to an object with mass, as many as $2^n-1$ virtual matter points are generated, acting as a perfect mass.

따라서 전자 하나를 가속하였을 때 대형 트럭과 맞먹는 충격량이 나올 수 있는 것이다. 그렇다고 전자 하나가 대형트럭 모양으로 변해서 나타나는 것은 아니다. (그만큼 많이 발생한 가상질점들의 숫자와 해당 질점들이 가진 관성질량이 큰 것이다.)

Therefore, when one electron is accelerated, an impulse equivalent to that of a large truck can be generated. However, this does not mean that one electron changes into the shape of a large truck. (The number of virtual matter points that occur that much and the inertial mass of those matter points are large.)

이 현상은, 가해지는 nF의 힘에 의해 누적되는 관성질량의 계속적인 누적 상태가 멈추고, 가속도의 증가가 멈추는 평형 상태가 이루어질 때까지 계속된다.

This phenomenon continues until an equilibrium state is reached where the continuous accumulation of inertial mass accumulated by the applied nF force stops and the increase in acceleration stops.

(그 평형 상태의 의미는 광속에 가까울수록 점점 질량 증가가 느려지다가 멈추고 광자의 속도 증가가 느려지다가 멈추는 평형을 이루는 지점에서 광속의 증가도 멈추는 현상을 통해 인류가 광속을 측정할 수 있게 된 것이라고 추정할 수 있다.)

(The meaning of the equilibrium state is that the increase in mass gradually slows and stops as it approaches the speed of light, and the increase in the speed of light stops at the point where the increase in the speed of photons slows and stops. It is assumed that mankind has been able to measure the speed of light. can do.)

그러나 유독 관성질량 증가와 속도의 증가인 가속도 관련하여 물리학에서는 평형 상태를 인정하지 않는다. (광속에 가까울수록 질량이 폭발적으로 증가한다고 말하고 있다.)

However, especially in relation to acceleration, which is an increase in inertial mass and speed, the state of equilibrium is not recognized in physics. (It is said that the closer you get to the speed of light, the more explosively the mass increases.)

다른 대부분의 물리현상에서는 평형 상태를 인정하는데 왜 질량 증가와 관련하여서는 평형 상태를 인정하지 않을까?

A state of equilibrium is recognized in most other physical phenomena, so why is a state of equilibrium not recognized in relation to mass increase?

이것이 질량 증가공식이 틀렸다는 증거가 된다.

This is proof that the mass increase formula is wrong.

우리가 모터를 다룰 때 쓰는 동력에 대해서 생각해보자.

Let's think about the power we use when operating a motor.

동력 = 일/시간 = 힘 × 거리/시간(단위: kg · m/sec, lb · ft/sec)

1HP(영국마력) = 76kg · m/s = 641kcal/h

1PS(국제마력) = 75kg · m/s = 632kcal/h

1KW = 102kg · m/s = 860kcal/h

우리가 만일 10마력의 모터를 가지고 가동을 한다면, 무부하 상태나 부하 상태나 항상 10마력의 힘을 소모하는 것이 아니다.

If we have a 10 horsepower motor running, it does not always consume 10 horsepower, either unloaded or loaded.

10마력의 모터로 750kg의 중심축이 있는 질량체를 회전시키려고 하면, 초기에는 많은 마력수가 필요하고, 10마력의 거의 근사치까지 에너지를 소모한다.

If you try to rotate a mass with a central axis of 750kg with a 10 horsepower motor, a lot of horsepower is initially required and energy is consumed up to approximately 10 horsepower.

그러나, 시간이 경과하여 모터가 설계된 지정속도에 이르기까지 회전을 지속하는 과정에서 모터의 소모전력은 급격히 줄어들면서 회전관성 모멘트가 최대가 된다.

However, as time passes and the motor continues to rotate up to the designed speed, the power consumption of the motor decreases rapidly and the moment of rotational inertia reaches its maximum.

즉, 10kw 발전기, 750kg의 플라이 휠, 그리고 10마력의 모터를 연결하였을 때와 플라이 휠 없이 발전기와 모터를 연결하였을 때 나오는 발전량은 현격한 차이가 있다. (물론 플라이 휠의 질량을 이루는 질점이 중심축에서 얼마나 멀리 떨어져 분포되었는지에 따라 결과에는 차이가 있다.)

In other words, there is a significant difference in the amount of power generated when a 10kw generator, a 750kg flywheel, and a 10 horsepower motor are connected and when the generator and motor are connected without a flywheel. (Of course, the results vary depending on how far the mass of the flywheel is distributed from the central axis.)

또 발전기는 5kw이고 모터가 10마력이고 플라이 휠이 750kg에 달할 경우에도 모터에서 소비하는 전력보다 발전기에서 나오는 전력량이 충분히 많을 수 있다.

Also, even if the generator is 5kw, the motor is 10

horsepower, and the flywheel weighs 750kg, the amount of power coming from the generator may be sufficiently greater than the power consumed by the motor.

이때, 발전기의 부하는 회전관성질량에 의하여 상쇄되고 그 회전관성질량의 설계에 따라 궁극적인 무부하 발전이 가능해진다.

At this time, the load of the generator is offset by the rotating inertial mass, and ultimate no-load power generation is possible depending on the design of the rotating inertial mass.

궁극적인 무부하 발전은 결국 회전관성질량을 전기에너지로 에너지화한 것이고, 이것은 결국 엔트로피 법칙의 붕괴를 증명하는 것이며, 엔트로피 법칙의 붕괴는 에너지 보존의 법칙 붕괴와 직결된다.

The ultimate no-load power generation is ultimately the energization of rotating inertial mass into electrical energy, which ultimately proves the collapse of the law of entropy, and the collapse of the law of entropy is directly related to the collapse of the law of conservation of energy.

일반적으로 모터의 출력량을 작은 양으로 계상하고 발전기의 출력량을 큰 것으로 만든 다음에 플라이 휠(flywheel)을 이용해서 에너지를 증폭하려는 오류를 범하기 쉽다.

In general, it is easy to make the mistake of calculating the motor's output as a small amount, making the generator' s output a large amount, and then trying to amplify the energy using the flywheel.

모터의 출력량은 발전기의 출력량과 상관없이 플라이 휠이 가지고 있는 질량을 얼마나 힘차게 돌릴 수 있느냐로 결정되어야 한다.

The output of the motor must be determined by how powerfully the mass of the flywheel can be turned, regardless of the output of the generator.

그리고 발전기는 얼마나 더 적은 손실로 발전할 수 있느냐를 별도로 생각해야 하는 것이다.

And the generator must consider separately how much less loss it can generate.

따라서 10마력의 모터와 750kg의 플라이 휠 그리고 10kw의 발전기를 연결하였을 때, 발전기에서 발생하는 회전을 방해하는 힘은 플라이 휠에서 발생하는 회전관성질량으로 충분히 상쇄될 수 있으며, 모터의 소비전력은 그만큼 최소가 된다.

Therefore, when a 10 horsepower motor, a 750kg flywheel, and a 10kw generator are connected, the force that interferes with rotation generated by the generator can be sufficiently offset by the rotational inertial mass generated by the flywheel, and the power consumption of the motor is becomes the minimum.

그러므로 엔트로피 법칙은 관성의 법칙과 대립되며, 관성의 법칙이 오류를 가지고 있을 가능성은 없다.

Therefore, the law of entropy is opposed to the law of inertia, and there is no possibility that the law of inertia contains an error.

그러므로 관성의 법칙은 엔트로피 법칙에 우선하는 자연계의 법칙이다. 따라서 엔트로피 법칙은 수정되어야 한다.

Therefore, the law of inertia is a law of the natural world that precedes the law of entropy. Therefore, the entropy law must be revised.

"인위적으로 관여하지 않는 한 엔트로피 법칙은 고-엔트로피로 가려는 경향이 있다. 이때 저-엔트로피와 고-엔트로피는 상대적이다."

"Unless artificially intervened, the law of entropy tends to go toward high entropy. At this time, low-entropy and high-entropy are relative."

"핵융합이나 핵분열의 경우에 의한 최종 물질이 질량이 있는 물체이고, 더 이상 핵분열이나 핵융합에 사용할 수 없는 고-엔트로피의 물질이라 할 수 있지만, 이 물질은 관성에 있어서는 항상 재활용이 가능한 저-엔트로피 물질인 것이다."

"In the case of nuclear fusion or nuclear fission, the final material is an object with mass and can be said to be a high-entropy material that can no longer be used for nuclear fission or fusion, but this material is a low-entropy material that can always be recycled in terms of inertia. "It is."

# 에너지 보존의 법칙
## Law of conservation of energy

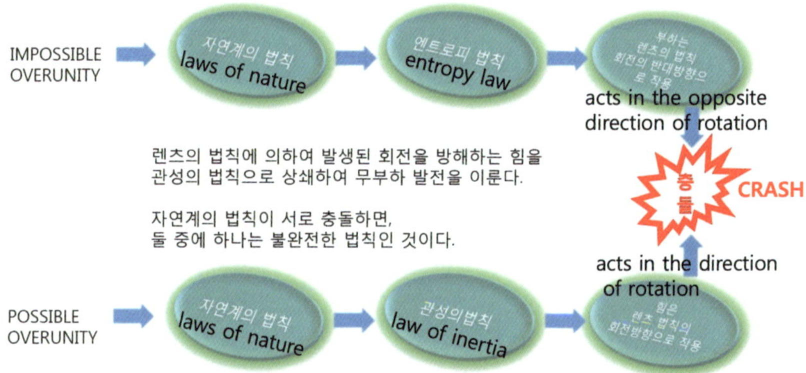


렌츠의 법칙에 의하여 발생된 회전을 방해하는 힘을 관성의 법칙에 기인한 관성질량 증가로 상쇄하여 무부하 발전을 이룬다.

No-load power generation is achieved by offsetting the force that interferes with rotation generated by Lenz's law with an increase in inertial mass due to the law of inertia.

자연계의 법칙이 충돌하면 둘 중에 하나는 불완전한 법칙인 것이다.

When the laws of the natural world conflict, one of the two

is an incomplete law.

우리는 지금까지 에너지 보존의 법칙이나 엔트로피 법칙이 자연계의 법칙으로 알고 그렇게 배워왔다.

Until now, we have been taught that the law of conservation of energy and the law of entropy are laws of the natural world.

외계에 접촉이 없을 때 고립된 계에서 에너지의 총합은 일정하다는 물리학의 법칙은 엔트로피 법칙이 붕괴되면 필연적으로 붕괴될 수밖에 없다.

The law of physics, which states that the total amount of energy in an isolated system is constant when there is no contact with the outside world, will inevitably collapse when the law of entropy collapses.

자연계의 법칙으로 알고 있는 법칙이 서로 충돌하면, 둘 중에 하나는 불완전한 법칙인 것이다.

If laws known as laws of the natural world conflict with each other, one of them is an incomplete law.

그런데 관성의 법칙이 오류가 있을 수는 없다. 다만 수식화되지 않았을 뿐이다.

However, the law of inertia cannot be in error. It's just that it hasn't been formalized.

우리는 언제부터인가 수학식으로 물리현상을 검증하려는 경향이 있다.

Since some time ago, we have been showing a tendency to verify physical phenomena using mathematical formulas.

## 관성은 어떻게 누적되는가?
## How does inertia accumulate?

그럼 지금부터 관성이 어떻게 누적되는지 살펴보자.

Now, let's take a look at how inertia accumulates.

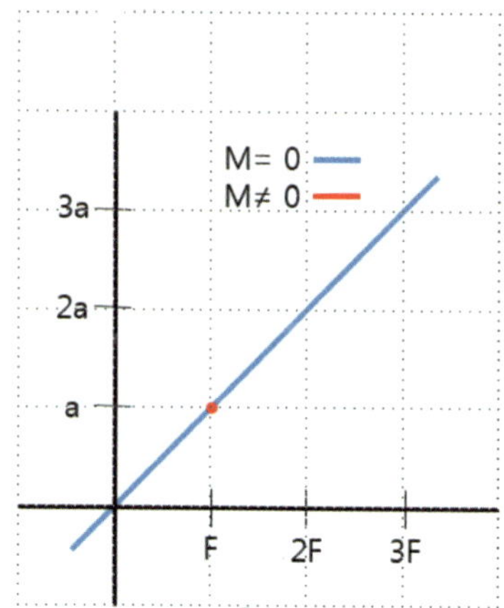


상기 그래프에서 직선 그래프는 질량이 0일 때 나오는 힘과 가속도의 관계를 나타낸 그동안 우리가 배워왔던 가속도 법칙에 부합하는 그래프이다.

In the graph above, the straight-line graph is a graph that conforms to the law of acceleration that we have learned, which shows the relationship between force and acceleration that occurs when the mass is 0.

그러나 질량이 0인 물체는 어떤 힘을 가해도 관성질량이 쌓이지 않는다. 다만 인간이 측정하기 어려운 아주 미약한 초기 질점(예를 들면 전자 1개의 질량)이 있다고 가정하자.

However, an object with a mass of 0 does not accumulate inertial mass no matter what force is applied. However, let us assume that there is a very weak initial mass (for example, the mass of one electron) that is difficult for humans to measure.

처음 가속도를 정하는 것은 힘 F에 의해서이다.

The initial acceleration is determined by force F.

그러므로 질량이 있는 물체의 가속도 또한 F의 힘에 의해 처음 a의 가속도를 내게 된다.

Therefore, the acceleration of an object with mass also produces an initial acceleration of a due to the force of F.

이때 질량 m에는 m'1라는 관성이 처음 생겨난다. (처음 F값이 0일 때는 정지질량 m의 값만 가지게 된다.)

At this time, an inertia called m'1 is first created in mass m. (When the first F value is 0, it only has the value of the rest mass m.)

m'1 관성은 m에 F의 힘을 가했을 때, 나타나는 두 가지 현상

중에 하나이다.

m'1 inertia is one of two phenomena that appear when a force of F is applied to m.

F의 힘이 가속도 a도 만들지만 m'1이라는 관성도 만든 것이다.

The force of F creates acceleration a, but also creates inertia m'1.

이때까지는 F=ma 공식이 완벽하게 성립한다.

Up to this point, the formula F=ma is perfectly established.

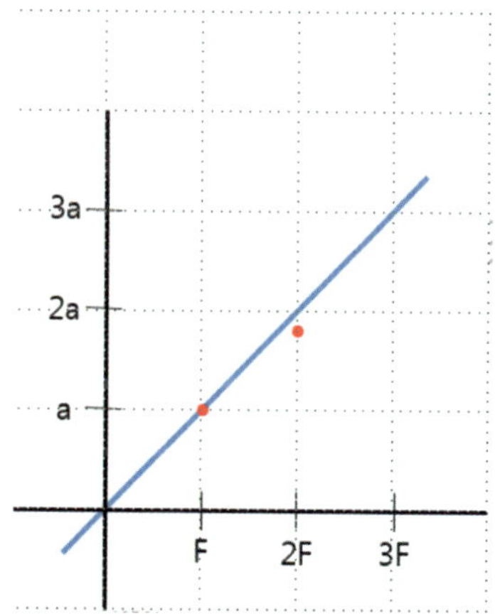


이제 2F의 힘이 가해졌을 때를 살펴보자.

Now let's look at when the force of 2F is applied.

이때에는 질량 m과 관성질량 m'1이 가해진 2F의 힘 중에서 일정량을 나누어 소비한다. 그 소비과정에서 m은 m'2를, m'1은 m'1-1의 관성질량을 또 만들어낸다.

At this time, a certain amount of the 2F force applied by the mass m and the inertial mass m'1 is divided and consumed. In the consumption process, m creates m'2, and m'1 creates an inertial mass of m'1-1.

따라서 2F의 힘이 가해 졌음에도 불구하고, 가속도는 2a에 도달하지 못하게 된다.

Therefore, even though a force of 2F is applied, the acceleration does not reach 2a.

만들어진 관성질량도 2F의 힘 중에서 일정 부분을 소비하기 때문이다.

This is because the created inertial mass also consumes a certain portion of the 2F force.

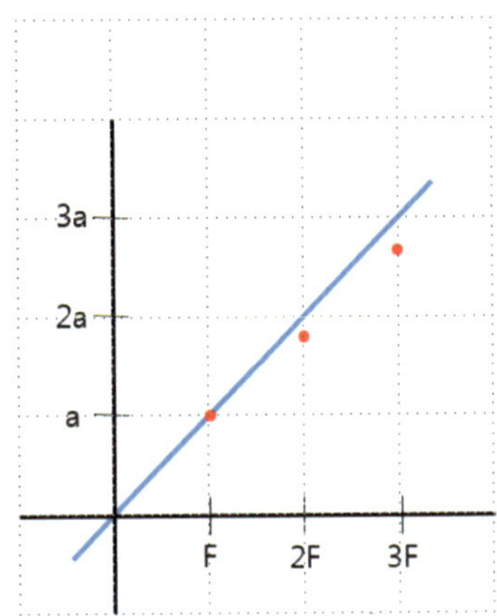


3F의 힘이 가해졌을 경우 질량은 m + m'1 + m'1 – 1 + m'2가 되고, 이 질량이 3F의 힘을 일정량씩 나누어 소비를 한다.

When a force of 3F is applied, the mass becomes m + m'1 + m'1 - 1 + m'2, and this mass consumes a certain amount of the force of 3F.

그리고, 그 결과물로 m은 m'3을, m'1은 m'1-2를, m'1-1은 m'1-1-1을 m'2는 m'2-1을 만들어낸다.

And, as a result, m produces m'3, m'1 produces m'1-2, m'1-1 produces m'1-1-1, and m'2 produces m'2-1.

따라서, 3F의 힘이 가해져도 가속도는 3a 도달하지 못한다

Therefore, even if a force of 3F is applied, the acceleration does not reach 3a.

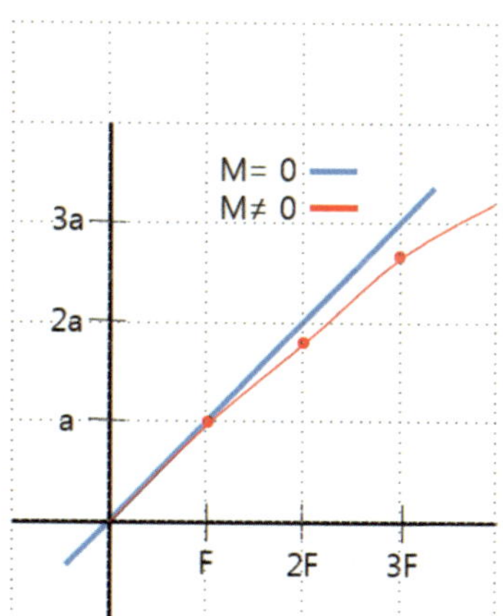


이렇게 배치된 빨간 점들을 연결하면 위의 그래프와 같이 누적된 관성력의 그래프 모양을 볼 수 있다.

By connecting the red dots placed in this way, you can see the shape of the graph of the accumulated inertial force, as shown in the graph above.

그래프는 완만한 곡선 형태를 띠게 된다. 이것을 표로 표현하면 다음 표와 같다.

The graph takes the form of a gentle curve. If this is expressed in a table, it is as shown in the table below.

| FORCE | GARVITY MASS | INERTAL MASS | | | | | RECORD |
|---|---|---|---|---|---|---|---|
| 0F | M | | | | | | 1 |
| 1F | M | m'1 | | | | | 2 |
| 2F | M | m'1 | m'1-1 | | | | 4 |
| | | m'2 | | | | | |
| 3F | M | m'1 | m'1-1 | m'1-1-1 | | | 8 |
| | | | m'1-2 | | | | |
| | | m'2 | m'2-1 | | | | |
| | | m'3 | | | | | |
| 4F | M | m'1 | m'1-1 | m'1-1-1 | m'1-1-1-1 | | 16 |
| | | | | m'1-1-2 | | | |
| | | | m'1-2 | m'1-2-1 | | | |
| | | | m'1-3 | | | | |
| | | m'2 | m'2-1 | m'2-1-1 | | | |
| | | | m'2-2 | | | | |
| | | m'3 | m'3-1 | | | | |
| | | m'4 | | | | | |

<table>
<tr><th>FORCE</th><th>GARVITY MASS</th><th colspan="5">INERTAL MASS</th><th>RECORD</th></tr>
<tr><td rowspan="16">5F</td><td rowspan="16">M</td><td rowspan="8">m'1</td><td rowspan="4">m'1-1</td><td rowspan="2">m'1-1-1</td><td>m'1-1-1-1</td><td>m'1-1-1-1-1</td><td rowspan="16">32</td></tr>
<tr><td>m'1-1-1-2</td><td></td></tr>
<tr><td>m'1-1-2</td><td>m'1-1-2-1</td><td></td></tr>
<tr><td>m'1-1-3</td><td></td><td></td></tr>
<tr><td rowspan="2">m'1-2</td><td>m'1-2-1</td><td>m'1-2-1-1</td><td></td></tr>
<tr><td>m'1-2-2</td><td></td><td></td></tr>
<tr><td>m'1-3</td><td>m'1-3-1</td><td></td><td></td></tr>
<tr><td>m'1-4</td><td></td><td></td><td></td></tr>
<tr><td rowspan="4">m'2</td><td rowspan="2">m'2-1</td><td>m'2-1-1</td><td>m'2-1-1-1</td><td></td></tr>
<tr><td>m'2-1-2</td><td></td><td></td></tr>
<tr><td>m'2-2</td><td>m'2-2-1</td><td></td><td></td></tr>
<tr><td>m'2-3</td><td></td><td></td><td></td></tr>
<tr><td rowspan="2">m'3</td><td>m'3-1</td><td>m'3-1-1</td><td></td><td></td></tr>
<tr><td>m'3-2</td><td></td><td></td><td></td></tr>
<tr><td>m'4</td><td>m'4-1</td><td></td><td></td><td></td></tr>
<tr><td>m'5</td><td></td><td></td><td></td><td></td></tr>
</table>

이때 F부터 5F까지 힘이 가해지는 동안, 질량 m에는 관성질량이 일정한 패턴을 그리며 쌓여간다.

At this time, while the force is applied from F to 5F, the inertial mass accumulates in a certain pattern in the mass m.

질량 m을 전자 1개 혹은 하나의 질점이라고 한다면,

If the mass m is one electron or one matter,

$1F = 2^1 = 2$

$2F = 2^2 = 4$

$3F = 2^3 = 8$

$4F = 2^4 = 16$

$5F = 2^5 = 32 \cdots$

$nF = 2^n$

즉, F의 힘을 가하기 전에는 질점은 움직이지 않는 m의 값을 갖는 1개의 질점이지만, nF의 힘이 차례대로 주어졌을 때, 질점은 $2^n$개가 되며, 초기 질점을 제외한 관성질점은 $2^n-1$개가 된다.

In other words, before applying the force of F, the material point is one material point with a value of m that does not move, but when the force of nF is given in order, there are 2n material points, and the number of inertial material points excluding the initial material point is $2^n-1$.

즉, 이 질점들은 질량에 비례하여 발생한 실체가 없는 가상의 질점이고, 그 각각의 값은 서로 다르다.

In other words, these matter points are intangible virtual matters that occur in proportion to mass, and their respective values are different.

그러나 이 가상의 질점들은 모두 완벽한 질량의 역할을 한다.

However, all of these virtual matter acts as a perfect mass.

**관성은 이런 형태로 누적된다. 그리고 그것은 결국 질량 증가다. 이러한 질량 증가는 로렌츠변환식으로는 증명할 수 없다.**

**Inertia accumulates in this way. And it is ultimately mass gain. This mass increase cannot be proven using the Lorentz transformation equation.**

질량은 관성의 수치적인 측정량이다. 질량의 단위는 SI 단위계에서 킬로그램(kg)이다.

Mass is a numerical measure of inertia. The unit of mass is kilogram (kg) in the SI unit system.

관성(질량)은 뉴턴역학에서 외부 힘에 대하여 저항하는 정도를 말한다.

Inertia (mass) refers to the degree of resistance to external forces in Newtonian mechanics.

정지한 물체에 힘이 가해지지 않으면, 그 물체는 정지를 계속한다. 운동하는 물체에 힘이 가해지지 않으면 그 물체는 운동 상태를 바꾸지 않고 등속 직선운동을 계속한다. 이것이 뉴턴의 제1 법칙이다. (출처: 위키백과)

If no force is applied to an object at rest, the object continues to rest. If no force is applied to a moving object, the object continues to move in a straight line at a constant speed without changing its state of motion. This is Newton's first law. (Source: Wikipedia)

그러므로, 아인슈타인이 주장한 질량 증가는 맞을지라도, 그 질량 증가가 광속에 가까운 상태에서 이루어지는 것이 아님을 알 수 있다.

Therefore, even though the mass increase claimed by Einstein is correct, it can be seen that the mass increase does not occur at a speed close to the speed of light.

오히려 질량 증가는 빠르게 가속도가 증가하는 구간에서 질량도 빠르게 증가한다는 것을 예측할 수 있다.

Rather, it can be predicted that mass increases rapidly in sections where acceleration increases rapidly.

따라서 아인슈타인이 로렌츠변환식을 차용하여 만든 질량 증가공식은 관성의 법칙을 위반하는 오류를 가지고 있음이 명백하다.

Therefore, it is clear that the mass increase formula created by Einstein by borrowing the Lorentz transformation formula has an error that violates the law of inertia.

그러나 애석하게도 저렇게 규칙적으로 관성이 쌓여가는 내용을 보고도, 필자 또한 질량 증가공식을 만들어내지는 못했다.

However, unfortunately, even after seeing the regular accumulation of inertia, the author was unable to create a formula for mass increase.

어쩌면 무수한 실험을 통해 상수를 만들어낼 수는 있을지도 모르겠다.

Perhaps it may be possible to create a constant through countless experiments.

분야가 다르지만 생화학에서 가장 잘 알려진 효소 반응속도론에 관한 모델 중 미카엘리스-먼텐 반응속도론에 의한 그래프와 유사한 그래프가 질량 증가 그래프이자 관성이 적용된 가속도의 법칙 그래프에서 보여지는 것이 합당할 것으로 생각한다.

Although the field is different, I think it would be reasonable for a graph similar to the Michaelis-Munten kinetics graph, among the most well-known enzyme kinetics models in biochemistry, to be shown as a mass increase graph and an acceleration law graph with inertia applied.

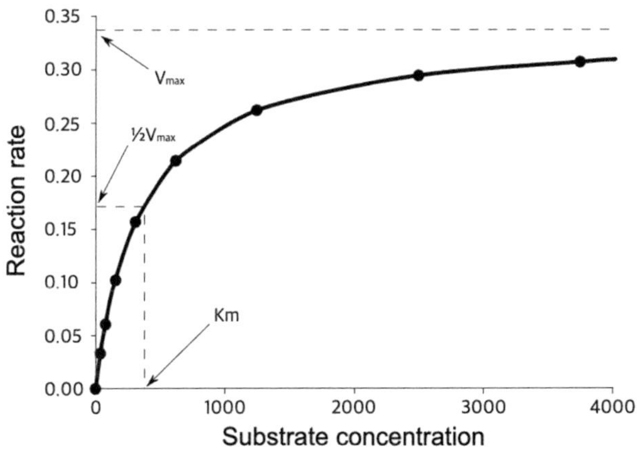


**미카엘리스-먼텐 동력학 평형 상태**

**Michaelis-Menten kinetics equilibrium state**

농도가 과량일 때, 일정한 한계속도에 접근하는 평형 상태는, 마치 질량의 증가가 한계일 때, 한계속도(광속)에 접근하는 것과 유사하기 때문이다.

This is because the equilibrium state approaching a certain limit speed when the concentration is excessive is similar to approaching the limit speed (the speed of light) when the increase in mass is limited.

이때 질량 증가는 광속에 가까울 때 폭발적으로 증가하지 않고, 오히려 초기에 증가하고 평형 상태에 가까울수록 증가량은 줄어든다고 볼 수 있다. (무거운 물체는 점점 더 던지기 어려운 법이니까…)

At this time, the mass increase does not increase explosively when it approaches the speed of light, but rather increases initially and the amount of increase decreases as it approaches the equilibrium state. (Because heavy objects become more and more difficult to throw…)

# 부록
supplement

# 렌츠의 법칙
## Lenz' s law

1834년, 독일계 러시아인 물리학자 하인리히 렌츠가 발견했으며, 어떤 폐회로에 유입되는 자기 선속(magnetic flux)이 변할 때, 유도되는 기전력은 그 자기 선속의 변화를 방해하게 만드는 자기장을 형성하게끔 생성된다는 법칙이다.

Discovered in 1834 by German-Russian physicist Heinrich Lenz, it is a law that states that when the magnetic flux flowing into a closed circuit changes, the induced electromotive force is created to form a magnetic field that interferes with the change in magnetic flux.

간단하게 설명하면, 자석과 가까워지면 코일은 자석을 밀어내려고 하고, 자석이 멀어지려고 하면 끌어당기려고 한다.

To put it simply, when the coil gets closer to the magnet, it tries to repel the magnet, and when the magnet moves away, it tries to attract it.

렌츠의 법칙에 의하여 맴돌이 전류가 발생하는데 이 현상을 잘못 확대 해석해서 렌츠의 법칙이 부하와 발열현상을 동반한다고 생각하는 경향이 있다.

Eddy currents are generated by Lenz's law, but there is a tendency to incorrectly interpret this phenomenon and think that Lenz's law is accompanied by load and heat generation.

부하는 회전을 방해하는 힘으로써 확실하게 렌츠의 법칙에 의한 현상이 맞다.

The load is a force that interferes with rotation and is clearly a phenomenon caused by Lenz's law.

그러나, 발열현상은 렌츠의 법칙에 의한 맴돌이 전류로 발생하는 발열현상과 렌츠의 법칙과 무관한 철심과 도체의 저항에 의하여 발생한다. (그래서 초전도체에서 저항이 0일 때 자기장의 움직임은 있지만, 저항 없이 전기 전송이 가능한 것이다.)

However, the heating phenomenon occurs due to eddy current according to Lenz's law and the resistance of the iron core and conductor unrelated to Lenz's law. (That is why, in a superconductor, when the resistance is 0, there is movement of the magnetic field, but electricity transmission is possible without resistance. (heat generation))

여기서 철심을 연자성 패라이트 고투자율 재료를 사용하면 철심에 의한 발열을 억제할 수 있다.

Here, if the iron core is made of a soft magnetic ferrite high permeability material, heat generation by the iron core can be suppressed.

남은 도체의 저항에 의한 발열은 구리선의 경우 발열이 일어나는 속도보다 방열되는 속도가 높을 경우 발열로 인한 열 손실을 걱정하지 않아도 된다.

In the case of copper wire, there is no need to worry about heat loss due to heat generation due to the resistance of the remaining conductor if the heat dissipation speed is higher than the heat generation speed.

발열은 저항에 의한 발열과 맴돌이 전류에 의한 발열이 발생할 수 있으나, 소재를 바꾸면 충분히 극복할 수 있는 문제다.

Heat generation can occur due to resistance or eddy shivering, but this is a problem that can be easily overcome by changing the material.

렌츠의 법칙을 해석할 때, 부하와 발열이 한 몸이라고 생각하지 않아야 한다.

When interpreting Lenz's law, one should not think of load and heat as one body.

- 회전을 방해하는 힘은 회전 운동을 유지하려고 하는 힘(관성력, 관성질량)을 이용하여 상계할 수 있다.
- The force that interferes with rotation can be offset by using the force (inertial force, inertial mass) that tries to maintain rotational motion.

- 발열현상은 소재를 바꾸어서 개선할 수 있다.
- Heating phenomenon can be improved by changing the material.

# 영점에너지
# Zero point energy

### 영점에너지의 의미
### Meaning of zero point energy

'영점에너지' 또는 '양자진공 영점에너지'는 양자역학계가 가질 수 있는 가장 낮은 에너지로, 그 계의 바닥 상태의 에너지를 말한다.

'Zero point energy' or 'quantum vacuum zero point energy' is the lowest energy that a quantum mechanical system can have and refers to the energy of the ground state of the system.

대단히 어려운 말이다.

This is a very difficult word.

이것을 쉽게 이해하고자 한다면, 영점에너지는 에너지의 이동통로 즉 에너지가 흐르는 가장 명확한 좌표이다.

To understand this easily, zero point energy is the clearest coordinate through which energy flows, that is, the path through which energy flows.

예를 들면, 온도에서는 절대영도가 에너지가 흐를 수 있는 마지노선 온도이고, 지구상의 좌표는 그 중심점이 에너지가 흐르는 좌표가 된다.

For example, in terms of temperature, absolute zero is the Maginot line temperature where energy can flow, and the center point of coordinates on Earth is the coordinate through which energy flows.

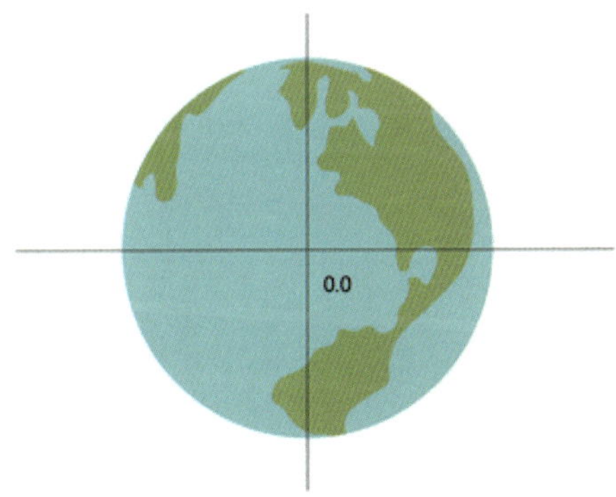


그러므로 중심점의 좌표는 (0.0)이다.

Therefore, the coordinates of the center point are (0.0).

이것이 영점에너지가 가지고 있는 의미이다.

This is the meaning of zero point energy.

따라서 모든 회전체에 있어서 에너지의 이동통로는 (0.0) 좌표를 관통하는 중심축이 된다.

Therefore, in all rotating bodies, the path through which energy moves is the central axis that passes through the (0.0) coordinate.

중심축을 벗어난 연결은 그 어떤 연결도 에너지의 손실을 야기시킨다.

Any connection that is off the central axis causes energy loss.

# 수소가 주는 교훈
## Lessons from hydrogen

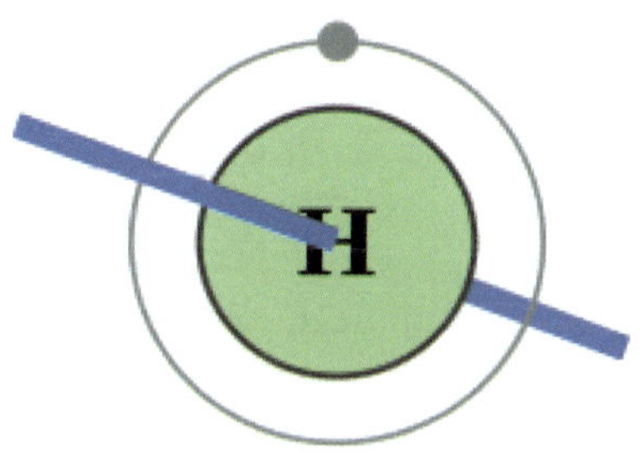


수소는 작은 회전체이자 전자를 회전시키는 질량체이다.

Hydrogen is a small rotating body and a mass that rotates electrons.

이 전자는 고정관념을 버리고 확대해석을 통해 풍력발전기의 블레이드와 같다고 생각해보자.

Let's think of these electrons as being like the blades of a wind turbine through expanded interpretation.

풍력발전기의 블레이드가 가벼운 경우 아무리 강한 바람이 불어도, 발전기의 부하를 이겨낼 수 없기 때문에 발전량이 줄어든다. (회전을 방해하는 힘은 회전방향과 반대방향으로 작동하는 렌츠의 힘이고 질량은 회전하는 방향으로 관성질량을 증가시킨다.)

If the blades of a wind generator are light, no matter how strong the wind blows, they cannot overcome the load on the generator, so the amount of power generated is reduced. (The force that interferes with rotation is the Lenz force that operates in the opposite direction to the direction of rotation, and the mass increases the inertial mass in the direction of rotation.)

지금까지 우리는 블레이드를 바람을 통해 회전시켜 전기에너지를 얻었다.

Until now, we have obtained electrical energy by rotating blades through the wind.

그러나 영점에너지의 의미를 확대하여, 블레이드의 앞 회전축에 모터를 달고, 뒤 회전축에 발전기를 달아서 바람 없이도 블레이드를 회전시킬 수 있다.

However, by expanding the meaning of zero-point energy, the blade can be rotated without wind by attaching a motor to the front rotation axis of the blade and a generator to the rear rotation axis.

이렇게 되면 블레이드는 결국 질량체인 플라이 휠로 대체될 수 있다.

In this case, the blades can eventually be replaced by the

mass body, the flywheel.

따라서 다음과 같은 장비의 개발이 필요하다.

Therefore, the development of the following equipment is necessary.

(1) 연자성 고투자율 페라이트 재료로 만들어진 발열현상이 거의 없는 발전기

(1) Generator with almost no heating phenomenon made of soft magnetic, high permeability ferrite material

(2) 연자성 고투자율 페라이트 코어로 만들어진 발열현상이 거의 없는 모터

(2) A motor with almost no heat generation made with a soft magnetic high permeability ferrite core

(3) 연자성 고투자율 페라이트 코어로 만들어진 변압기

(3) Transformer made of soft magnetic high permeability. ferrite core

(4) 블레이드를 대체할 플라이 휠

(4) Flywheel to replace blade

이제 인류는 증가된 관성질량에서 에너지를 꺼내어 사용할 수 있을까?

Can humanity now extract energy from the increased

inertial mass and use it?

## 광자는 질량이 있는가?
## Do photons have mass?

인간의 오랜 궁금증 중에 하나이다.

It is one of humanity's long-standing curiosity.

그러나, 결론은 아주 간단하다.

However, the conclusion is very simple.

질량에너지 등가관계가 성립함에 따라 에너지를 가지고 있는 광자는 질량을 가지고 있다.

As the mass-energy equivalence relationship is established, photons with energy have mass.

광자가 질량이 0이라고 한다면, 광자는 에너지를 가질 수 없다.

If a photon has zero mass, then the photon cannot have energy.

그러므로 광자는 질량을 가지고 있다.

Therefore, photons have mass.

다만, 에너지가 있기에 질량을 가지게 된 것인지 아니면 질량이 있는 입자에 에너지가 추가되어 광자가 된 것인지는 아직 모른다.

However, it is not yet known whether it has mass because it has energy, or whether energy is added to a particle with mass to become a photon.

또 진동수(흔들림)와 파장을 가지는 파동도 물질이 전파되는 현상을 말하므로 질량을 가진 광자가 이동하는 방식 중에 하나인지 아니면, 파동도 광자가 질량을 가지게 만든 원인인지는 알지 못한다.

Also, since waves with frequency (vibration) and wavelength refer to the phenomenon in which matter propagates, it is not known whether this is one of the ways in which photons with mass move, or whether waves are also the cause of photons having mass.

광자를 움직이지 못하게 정지시키고, 에너지를 빼앗은 후 정지 질량을 측정하기 전까지는 알 수 없는 것이다.

This cannot be known until the photon is stopped from moving, its energy is taken away, and the rest mass is measured.

광자 1개의 에너지를 측정해서 1개의 질량값으로 환산할 수는 있을지 모르겠다.

I don't know if it is possible to measure the energy of one photon and convert it to one mass value.

## 힘의 단위는 사용할 수 있는 단위인가?
## Is the unit of force usable?

힘의 단위는 뉴턴(newton, 기호: N)이다.

The unit of force is newton (symbol: N).

아이작 뉴턴(1642. 12. 25.-1726. 3. 20.)이 사망한 이후 1960년에 도입되었다.

It was introduced in 1960 after the death of Isaac Newton (December 25, 1642 - March 20, 1726).

뉴턴은 1kg의 질량을 갖는 물체를 1미터에 매초제곱(1m/$S^2$)만큼 가속시키는 데 필요한 힘으로 정의된다.

Newton is defined as the force required to accelerate an object with a mass of 1 kg by 1 meter per second squared (1 m/$S^2$).

$1N = 1kg \cdot m/s^2$

그러나, 1N을 F라고 가정하고, 질량 m=1kg, 가속도 a=1 m/$s^2$이라고 가정했을 때, 처음 F의 힘이 가해졌을 때, 가속도 a까지는 정확하게 맞을 수 있다. 그러나 2F, 3F의 힘이 주어졌을 때, 가속도는 2a, 3a에 도달하지 못하게 된다. 질량이 계속 증가하기 때문이다.

However, assuming 1N is F, mass m=1kg, and acceleration a=1 $m/s^2$, when the force of F is first applied, the acceleration a can be accurately achieved. However, when forces of 2F and 3F are given, the acceleration does not reach 2a or 3a. This is because the mass continues to increase.

질량이 일정할 때라고 가정하는 것은 이렇게 많은 오류를 범하게 만든다.

Assuming that the mass is constant leads to many errors.

# 에필로그
# Epilogue

인간은 에너지에서 자유로워질 수 있는가?

Can humans become free from energy?

필자도 그것이 궁금하다.

I'm curious about that too.

지금까지 인류는 에너지 보존의 법칙에 갇혀 있었다.

Until now, humanity has been confined by the law of conservation of energy.

그리고 인류는 아인슈타인에 의해 질량-에너지 등가원리를 받아들였다.

And mankind accepted the mass-energy equivalence principle through Einstein.

**질량-에너지 등가(Mass-Energy Equivalence):**

모든 질량은 그에 상당하는 에너지를 가지고 있고, 그 역 또한 성립한다. (모든 에너지는 그에 상당하는 질량을 가진다.)

Every mass has a corresponding energy, and vice versa. (All energy has a corresponding mass)

또한 특수상대성 이론을 통하여 질량이 증가한다는 것을 받아들였다. (물론 공식의 오류가 있었어도 말이다.)

Additionally, it was accepted that mass increases through the special theory of relativity. (Of course, even if there was an error in the formula.)

질량이 증가한다는 것은 그만큼 에너지가 증가했다는 것을 뜻한다. (질량에너지 등가의 원리를 받아들이면 그렇다.)

An increase in mass means an increase in energy. (Yes, if you accept the principle of mass-energy equivalence.)

이제 이 책 속의 내용을 통해 질량이 광속에 가까울수록 기하급수적으로 증가한다는 허구에서 빠져나오게 될 것을 믿는다. (특히 질량 증가는 초기 누적된 관성질량이 적을 때, 질량에 비례하여 많이 일어난다.)

Now, I believe that through the contents of this book, we will be able to escape from the fiction that mass increases exponentially as it approaches the speed of light. (In particular, mass increase occurs in proportion to mass when the initial accumulated inertial mass is small.)

질량이 증가하는데 왜 에너지를 꺼내 쓰지 못하겠는가?

As mass increases, why can't energy be extracted and used?

과제는 증가한 질량만큼을 인간이 사용할 수 있는 에너지로 변환하는 것이다. (그것도 효율적으로…)

The challenge is to convert the increased mass into energy that humans can use. (And efficiently…)

이것이 에너지 자유를 위한 첫걸음이라 생각한다.

I believe this is the first step toward energy freedom.

이 책은 인류에게 큰 고통을 주고 있는 아토피 피부염의 원인을 밝혀준다.

This book reveals the cause of atopic dermatitis, which is causing great suffering to humanity.

물론 치료방법도 공개하였다. (전문가들이 나서서 더 많이 연구를 진행해야 한다.)

Of course, the treatment method was also disclosed. (experts need to step forward and conduct more research.)

그리고 인류가 자유롭게 사용할 수 있는 새로운 막수송체 기술도 공개하였다.

Additionally, a new membrane transporter technology that can be freely used by mankind was also revealed.

믿는 방법은 간단하다.
The way to believe is simple.

실천해서 사실인지 확인하면 된다.
All you have to do is practice it and see if it is true.

이 책을 끝까지 읽어주신 모든 독자분들께 감사드린다.
Thank you to all readers who read this book to the end.

2024년 3월

저자 **윤종오**